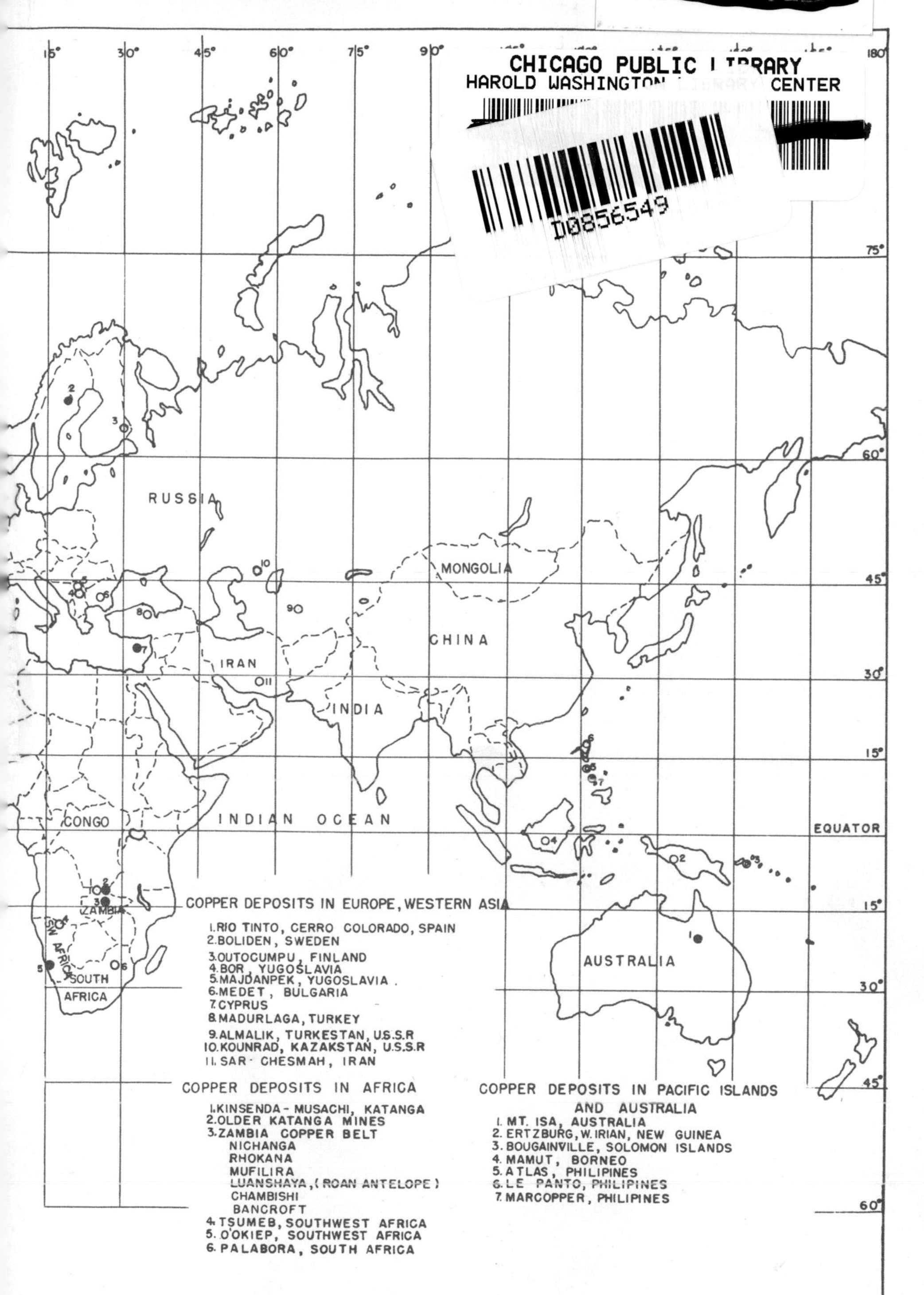
15° 30° 45° 60° 75° 90° 180°
75°
60°
45°
30°
15°
EQUATOR
15°
30°
45°
60°
RUSSIA
MONGOLIA
CHINA
IRAN
INDIA
INDIAN OCEAN
CONGO
ZAMBIA
SW AFRICA
SOUTH AFRICA
AUSTRALIA
COPPER DEPOSITS IN EUROPE, WESTERN ASIA
1. RIO TINTO, CERRO COLORADO, SPAIN
2. BOLIDEN, SWEDEN
3. OUTOCUMPU, FINLAND
4. BOR, YUGOSLAVIA
5. MAJDANPEK, YUGOSLAVIA
6. MEDET, BULGARIA
7. CYPRUS
8. MADURLAGA, TURKEY
9. ALMALIK, TURKESTAN, U.S.S.R
10. KOUNRAD, KAZAKSTAN, U.S.S.R
11. SAR- CHESMAH, IRAN
COPPER DEPOSITS IN AFRICA
1. KINSENDA - MUSACHI, KATANGA
2. OLDER KATANGA MINES
3. ZAMBIA COPPER BELT
NICHANGA
RHOKANA
MUFILIRA
LUANSHAYA, (ROAN ANTELOPE)
CHAMBISHI
BANCROFT
4. TSUMEB, SOUTHWEST AFRICA
5. O'OKIEP, SOUTHWEST AFRICA
6. PALABORA, SOUTH AFRICA
COPPER DEPOSITS IN PACIFIC ISLANDS AND AUSTRALIA
1. MT. ISA, AUSTRALIA
2. ERTZBURG, W. IRIAN, NEW GUINEA
3. BOUGAINVILLE, SOLOMON ISLANDS
4. MAMUT, BORNEO
5. ATLAS, PHILIPINES
6. LE PANTO, PHILIPINES
7. MARCOPPER, PHILIPINES

COPPER

This is one of a pair of hoisting engines, the Modoc and the Aztec, that were built in San Francisco for the Anaconda Mining Company by the Union Iron Works in 1896. Used at Butte, Montana, they were the largest hoisting engines in the world, capable of raising four-deck cages at the rate of 2400 feet per minute. (—MINING AND SCIENTIFIC PRESS)

COPPER

The Encompassing Story of
Mankind's First Metal

by Ira B. Joralemon

BERKELEY

CALIFORNIA

COPPER

Printed and bound in the United States of America

Library of Congress Catalog Card No. 73-88141

ISBN 0-8310-7103-6

Published by Howell-North Books
1050 Parker Street, Berkeley, California 94710

A PROSPECTOR'S DEDICATION

TO MY WIFE

I got a good woman, too. She's a hard worker, and a good cook, and she ain't so good-lookin' so that any other fellar besides myself is likely to get stuck on her.

Preface

It is hard to realize that nearly forty years have slipped by since I wrote the stories that were published as *Romantic Copper*. Though they have passed so quickly, the years have been eventful for copper. In those forty years nearly as much copper has been found or made available by technical progress as all that had been mined since our savage ancestors found the first metallic pellets in the campfires on hearths made of pretty green rocks.

The newer discoveries have not been as dramatic as those of earlier times. The luck or subconscious reasoning of the prospector has not played as vital a role. Still, chance and daring and the old-fashioned "nose for ore" continue to play a part. Without them this would be a dull story.

The change since the great Depression of 1929 to 1933 has been so startling that the story of copper divides itself into two parts. The first part includes tales of the great discoveries in all the centuries before 1930. I am reprinting these as they were told in *Romantic Copper* in 1933. Any attempt to rewrite them would take away the excitement of an era in which I was lucky enough to participate. In the first thirty years of this century the copper industry was so small that most engineers for the copper companies knew each other well. I worked in most of the American copper districts that were great before 1930. In some, such as Bisbee, New Cornelia, and Jerome, I took an active part in finding the ore; in others I made estimates for use in financing or in connection with tax problems. I also overlapped the western pioneers, who told me their stories. Much of *Romantic Copper* was based on talks with engineering or pioneer friends.

This personal approach resulted in a few protests from those who thought statements were prejudiced. As it is impossible to say just where facts lay half a century ago, I am quoting the principal protests at the ends of chapters or sections. Readers can take their choice which version to believe.

In the years since 1930 copper has changed from a small and intimate industry to a great one. The individual counts for much less, and so many engineers and geologists have taken part in the new developments that they know each other casually, if at all. My own connection with eight or ten of the new mines has been chiefly in examinations and estimates that were used in financing or in tax arguments.

With the unbelievable growth of the industry there has been a change in the way new mines are found. In earlier years imagination and vision that were often subconscious reasoning played a major role. Then all sorts of people shared in the fun and in the rewards. Prospectors, promoters, hard-boiled miners, and adventurous pioneers made most of the discoveries; trained engineers and geologists often played only a minor part. It was truly the age of romance in copper, for true romance is fact seen through rosy tinted glasses.

Since the ten-year hiatus in discoveries due to the great Depression, this sort of romance has gone out of style. The discovery of new types of ore bodies has still been due in many cases to prospectors and other adventurers in ideas, but most companies just tried to find new ore bodies like the ones others had found. And they succeeded. Most of the success was due to detailed study, aided by geophysics and geochemistry and all of the other new technical tools. It was effective, but it wasn't as much fun as the old exploration. And the new methods of transportation — the airplane and helicopter and jeep — weren't as satisfying, and didn't lead to thought, as much as the slower travel by ship and train and muleback.

The change in the copper industry since 1930 makes it inevitable that the second part of my book is more technical, and so less exciting in many ways, than the first part. I have tried to avoid too many statistics. Perhaps I have exaggerated the part played by those who have led the way in finding new types of ore bodies, but I think they are the ones most responsible for the progress.

One thing I don't want to change is the dedication that started *Romantic Copper*. This I borrowed from my old friend, pioneer, prospector and "raconteur par excellence," Jim Finch. More than fifty years ago Jim came down to the railway station a few miles from Lake Valley, New Mexico, to meet John Greenway and me. We wanted to buy some dumps Jim controlled that might make good smelter flux. Jim greeted us with a flood of words: "By God, gentlemen, I've been the busiest man in Sierra County this month. Bought a quarter interest in the Martha Washington claim, a half interest in the Salamander, bought a chicken ranch, and got married. I got a good woman, too. A hard worker and a good cook and she ain't so good-lookin' that any other feller besides myself is likely to get stuck on her." This was the highest praise Jim could think of for his wife. It was justified, too, as we found out from the meals we "ate behind" Mrs. Finch in the days we spent at Lake Valley.

I had told the story to my wife, and when she knew I was going to dedicate *Romantic Copper* to her, she insisted that I use Jim's words. Those who knew her would see the joke. So in it went. I use it again because, like Jim Finch, I want to express my highest appreciation.

IRA B. JORALEMON

San Francisco
March, 1973

Acknowledgments

THE principal source of my material for *Copper* has been my close association with copper mines and those who developed them. Among the thousand mines I examined were more than thirty of the great copper mines and districts. Examinations of several of them started at the beginning of their history and lasted for many years. The managers and staffs became my good friends, and they talked freely about their problems and hopes. Talks with engineers at technical society meetings were almost as helpful. The present book is in many ways a memorial to a host of friends in the industry. Sad to say, few of them are still alive.

My appreciation of the friendship and help of a few is particularly great. They include: John C. Greenway of Calumet and Arizona and New Cornelia; Dr. L. D. Ricketts of Greene Cananea, New Cornelia, Inspiration, Anaconda, etc.; B. B. Thayer, Reno H. Sales and William Braden of Anaconda and its subsidiaries in South America; James S. Douglas of United Verde Extension and his son Lewis W. Douglas; Louis S. Cates of Ray Consolidated and Phelps Dodge; Harry Lavender of Calumet and Arizona, New Cornelia and Phelps Dodge; Col. Seeley W. Mudd and his sons Harvey and Seeley Jr. of Cyprus Mines; David D. Irwin, William Burns and Thorold Field, old Bisbee friends who played a large part in the development of the Northern Rhodesia copper mines and whom I often talked with in later years; Neil Lakenan of Nevada Mines Division, Kennecott Copper Corporation; Robert Marsh of Consolidated Coppermines Corporation, later absorbed by Nevada Mines; Thayer Lindsley of Falconbridge Nickel Mines and many other companies; and Henry deWitt Smith of Kennecott, United Verde and Newmont.

In addition to these friends and associates, I have received friendly letters, historical material and pictures of mines and plants from Spud Huestis, who was largely responsible for Bethlehem Copper in British Columbia, and from Mr. R. F. StG. Lethbridge of Henley-on-Thames, England, and Mr. Leonard U. Salkield of Punta Chulera, Spain, both of whom have been directors or officers of some of the Rio Tinto group of companies. I am grateful for their help.

My knowledge of the great copper mines due to personal association has been supplemented by many years of study of annual and other reports of the companies and by reading, and summaries of, articles in technical journals.

I also wish to express my appreciation to the following persons and organizations who have sent pictures or historical data for *Copper:* George O. Argall, Jr., Editor, *World Mining;* J. C. Courage, Public Relations Department, Charter Consolidated Ltd.; Edwin E. Dowell, Manager of Public Relations, and R. F. Alkire, Public Relations Director (Nevada Mines Division), Kennecott Copper Corporation; J. W. Hanley, Director of Metallurgy, Cerro Mining Company; Don H. Hoskins, Director of Public Affairs, and Derek J. Wing, Chief Photographer, International Nickel Company of Canada, Ltd.; K. A. Howard, Copperbelt Public Relations Officer, Roan Consolidated Mines Ltd.; W. O. Irish, Director of Administration, Cyprus Mines Corporation; H. Myles Jacob, President, Inspiration Consolidated Copper Co.; W. W. Little, Assistant General Manager, Phelps Dodge Corp., Douglas, Ariz.; Dean W. Lynch, Public Relations Representative, Duval Corporation; F. Hugh Magee, Assistant to the Director of Public Relations, Freeport Minerals Company; Robert H. Ramsey, Newmont Mining Corporation; Mrs. Harrison Schmitt, whose husband was one of the Duval "trio"; A. W. Smith, Public Relations Officer, Selection Trust Ltd.; C. L. Sonnichsen, Chief of Publications, and Heather S. Hatch, Assistant Librarian, Arizona Historical Society; W. D. Thompson, Manager, Corporate Communications, Placer Development Ltd.; Irene Waisanen, Editor, *Daily Mining Gazette;* The Bancroft Library, University of California, Berkeley; the Engineering Societies Library; the Michigan Technological University Library; and the University of Nevada, Reno, Special Collections Library.

IRA B. JORALEMON

Contents

PART I

Foreword

IN THE long story of copper there have been many adventures that never are told in the technical histories. For technical histories must be solemn. In them the engineer must surmount all obstacles with unerring skill. The accidents through which Fate has often brought success to those least qualified to win it might detract from the dignity of a great profession if they were too widely advertised. If the engineer can not be given the glory, at least history can keep silent.

The unwritten stories have been told and retold over the campfires and poker tables of the mining country until they have almost become folklore. Like all folklore, they appeal so strongly to the spirit of romance that they have often grown in the telling. A germ of truth is there. But it is often impossible to sort out the truth from the fancy that has colored it.

It is hard to say just what truth is, anyhow. Men change so much in the course of their adventurous lives that what is true of one period is often entirely foreign to the next one. The Bill Greene who gravely discussed a concentrator flow sheet with Dr. Ricketts was quite a different person from the Bill Greene who used to get drunk in the Tombstone gambling halls with Jim Finch. Dr. Ricketts can't possibly believe the stories of Bill's early life, even if they are true. Bill Greene himself probably forgot a lot of the more lurid episodes.

History isn't quite fair about the way it treats its heroes. It accepts as gospel truth the stories of the respectable associates of their days of glory. And the greater the success the more credit the hangers on give the wisdom and virtue of their idols. But if the hero in his younger days ever had anything to do with a swindler or a cut-throat, the things the swindler or cut-throat say about him are thrown out of court. History forgets that if a man spends ten years surrounded by liars, it is going to be hard to get a picture of those ten years without accepting a liar's testimony.

It is not quite so bad as that in the case of the Copper Kings. But the recorded truth is usually pretty well concealed by what ought to have been true.

Fate seldom picked the respected and studious engineer as the discoverer of the great ore bodies. She let her favors drop indiscriminately on the ignorant prospector, the engineer, the gambler and the stock swindler. Even a pig claims to have found the Calumet and Hecla — but has no creditable witnesses to prove his claim. When it is an engineer who draws the winning cards, it is usually one who likes to go out on a party with the other boys and girls instead of staying home to write his memoirs so that history can get it all straight.

The only men who knew just what happened were those who were there when the copper discoveries were made. And they are most of them dead. They did not often get a chance to put their version of the affair in print — if they could write at all. But they told a lot of good stories about it, and they believed the stories. The tales may have grown a bit from year to year. They certainly do not agree with the histories written after Fame had descended on her favorites. But they do at least give the real atmosphere in which the events occurred. And in that they often have a deeper truth than the signed and attested versions.

The sketches that follow attempt to give some of the copper history that does not get in the official obituaries. They have all been told as true by men who ought to have known the facts. True or not, they try to show how the heroes' early associates felt about them. And if they are lucky, the stories may give out a little of the glamour of the unexpected that has always attended the search for copper.

1
The Island of Copper

MANY thousand years ago, long before the dawn of history, the half-savage tribe that lived in the island of Cyprus found some heavy red stones. They weren't brittle, like the rock they were accustomed to use for axes and spear-heads. Instead they bent when they were hammered. This was a wonderful quality, for it allowed the ancient workmen to beat the rocks into knives or axes or any other form they desired. The people who had these red stones looked down on their less fortunate neighbors who only had the common, brittle rocks. The new material was so useful that it became their choicest possession. To honor it, they called it from the name of their island home – the "Cyprian metal." It has kept the name through all the ages. Our tongues have changed it to "copper."

For at least thirty centuries the island of Cyprus was the most abundant source of the new metal. When the boulders of pure copper could no longer be found, the growing intelligence of the Cyprians, aided by accident, taught them to burn certain brilliant green rocks they found near by with charcoal. In the ashes they found lumps of copper – for the green rocks were copper carbonate. The Ionians, the Greeks, and finally the Romans worked these copper mines of Cyprus. Then for some strange reason the industry came to an end. For almost two thousand years the world looked elsewhere for its copper. The mines of Cyprus were forgotten. Everyone knew the name "copper" came from "Cyprus," but no one knew or cared why.

About the beginning of the twentieth century the rapid growth of the Electrical Age threatened an acute shortage of copper. Engineers and prospectors by the thousand started to search for new deposits. The most likely place to hunt surely ought to have been the island that gave copper its name. This conclusion was so very obvious that no one thought of it at all. Instead, the trail back to the oldest of copper mines was long and devious. It started in a grimy cañon in Utah; wound for a year through the Mexican desert; paused to wait while a girl made herself a little lovelier than usual for a theater party; and finally ended with the failure of a scientific search and the last moment discovery of an entirely different sort of deposit from the one that was hoped for. If it had not been for the repeated interference of blind Chance, the Cyprus copper deposits would still be a half-forgotten legend.

D. C. Jackling unwittingly started the chain of events that ended in Cyprus. While still an unknown young metallurgist, he conceived the idea that if he did on a gigantic scale the things that others had tried unsuccessfully to do on a small scale, he could make copper cheaply out of the great mass of lean copper-bearing rock that Colonel Wall had developed in Bingham Cañon, Utah. Most engineers laughed at the idea. Material that contained only 1.5 per cent copper, scattered or "disseminated" in minute grains of sulphide through solid rock, could not possibly yield a profit. Luckily for the world, they were wrong and Jackling was right. By 1907 his five-thousand-ton mill was running satisfactorily. The idea that seemed so simple made Utah Copper the greatest, and the lowest-cost copper mine in the country.

The success of Utah Copper was followed by an intensive search for similar disseminated copper deposits. Ore bodies of this type were easy to find. The oxidation or "rusting" at the surface of the iron sulphide that accompanied the copper sulphide caused brilliant red, yellow or brown caps or "outcrops" above the ore. Guided by this staining of the surface rock, Jackling and his associates and followers quickly developed nearly a dozen of the new low-grade ore bodies. The one simple idea made an entirely new crop of eminent engineers and of copper kings and doubled the world's supply of copper.

All of the disseminated deposits were found within half a dozen years.* But the active search went on for another decade. While it failed to find additional disseminated ore bodies, it had many interesting by-products. One of them was Cyprus.

Among the many engineers who took an active part in the hunt for new low-grade copper mines were Seeley W. Mudd and Philip Wiseman. They had been largely responsible for the successful development of the Ray Consolidated Copper Company in Arizona – one of the great Jackling companies. A thoughtful study of all the new disseminated ore bodies convinced them that desert climate was necessary for the formation of these deposits. The alternation of periods of dryness with torrential rains caused the concentration of the extremely lean material into flat blankets of 1 to 3 per cent ore that could be readily mined and concentrated at low cost. The desert should hide other ore bodies like those already developed.

A young engineer named D. A. Gunther was selected to do the scouting for the Mudd-Wiseman exploration syndicate. The southwestern United States and northern Mexico were his field, another Utah Copper his goal. For over a year he followed the desert trails, running down every wild story of red mountains that might contain the ore bodies he was looking for. But even his keen imagination could not see any hope of success. A few drill holes, sunk to test long chances that proved futile, ended the campaign. A disappointed engineer went back to his family home in Brooklyn to rest from the hardship in the desert, and to hunt for another exploring job.

After the lonely months in the wilderness, New York seemed very good to Gunther. He had scarcely spoken to a cultivated girl in two years, and his only theater had been an occasional dingy road show in a desolate border town. So he enjoyed himself to the full. Sometimes he could combine the com-

*This was true until the great discoveries of low-grade ore from 1935 to 1970.

panionship of an attractive girl and a good play, and so double the happiness. As he was often in Manhattan during the day, calling on engineering acquaintances in the hope that some job might develop, he used to meet the young lady who was then the center of his attentions at the Public Library before going to the theater.

Any girl would want to look her best in the company of an attractive young engineer just back from romantic wanderings in the desert. One evening the girl of the hour was delayed. Gunther passed the time looking over the books on the shelves near the meeting-place. He had an inquiring mind, and out of curiosity picked out the books that looked as though they had never been read. One of them was a treatise on archaeology. As he turned the pages, a reference to the use of copper by the Phoenicians caught his eye — and his mind. Where had these ancients found their copper? He had a vague recollection that the Rio Tinto deposits in Spain had been mined for thousands of years. He had seen bits of copper slag from Asia Minor in college museums, and recalled the long-forgotten fact that the name Cyprus came from the Greek word for copper, or vice versa. Curious, he thought, that the ancient mines had been in areas with climates not unlike that of our Southwest. Maybe conditions around the Mediterranean had been favorable for great enrichment of copper ores near the surface, like that in Bisbee or Clifton, in the Arizona desert. But if there were high-grade enriched ores there might also be low-grade disseminated or porphyry ores like those at Bingham or Miami or Ray. All of the southwestern disseminated ore bodies had been found in districts where smaller high-grade ore bodies had been mined for many years. Disseminated ores would have been worthless in ancient times, as it had been impossible to recover the copper from them without excessive expense until Jackling added large-scale operations to modern metallurgy. With the climate and topography the same, and copper ore bodies rich enough to be mined and treated before the dawn of history, somewhere around the Mediterranean there ought to be one of the great low-grade "porphyry" ore bodies he had been looking for so unsuccessfully along the Mexican border.

Before his belated friend arrived, Gunther's plan of campaign was made. It was the ideal prospecting venture. Added to the excitement of finding a great new mine was the lure of distant places and of the names that had seemed in another world when he had read them in school. He probably had no idea what play he was seeing, and the girl found him a most unattentive swain that evening.

The next day he wrote to Mr. Mudd and Mr. Wiseman about his great idea. It appealed to them at once. The wonder was that they had not thought of it before. Their syndicate agreed to finance the venture, and Gunther started at once to gather all the information he could find about the copper industry of ancient times.

Translations of Greek and Latin manuscripts in the New York Public Library and the British Museum suggested many possibilities. The desert surrounding Mount Sinai seemed to present the best chances. The Mesopotamian civilizations, as well as the Greek and Roman, had won much copper from this part of Asia Minor, and climatic conditions were much like those in Arizona. Rio Tinto and neighboring parts of southern Spain had been the greatest source of copper for three thousand years or more. But the great lenses of solid sulphide in Spain had been so thoroughly explored in recent years that it seemed unlikely that disseminated ore bodies could have been overlooked. Manuscripts occasionally mentioned Roman copper mines in the northern part of Africa. There desert conditions were certainly pronounced, and further investigation seemed justified.

Cyprus was mentioned as an ancient source of copper in many old writings. Ever since 3000 B.C. this island had been noted for its mines of the red metal that it had named. The Romans, the Greeks, the Phoenicians, and even the Ionians, with their highly developed civilization before the dawn of history, had looked on Cyprus as one of the chief sources of the metal that distinguished them from their Stone Age enemies.

Pliny, Aristotle, Herodotus and other early writers mentioned the copper of Cyprus as so well known that it hardly needed description. Strange to say, a more definite account of these Cyprian mines proved elusive. It was only by accident that Mr. Mudd found that Dr. Joseph Walsh, of Philadelphia, had come upon rather detailed references to the Roman mines in Cyprus in translating the works of Galen, physician to Marcus Aurelius in the second century, A.D.

Mr. Mudd wrote to Dr. Walsh and obtained from him a translation of the passages in Galen telling of a visit to the Skouriotissa Mine, in the year 166. At that time the ore that could be smelted into copper was apparently exhausted. The only production was by running drifts or inclines just below water level and allowing the ground water, heavily charged with copper sulphate, to drop into jars. Slaves carried the jars to the surface. Then the water was evaporated and copper sulphate, or "chalcanthus," crystallized. This was the basis for medicines, pigments, and other copper products. Unsatisfactory as the description was, it at least gave one definite point in Cyprus where copper had been found by the ancients.

A long campaign of field work was the next step in the venture. Asia Minor was the first area chosen. There Gunther assembled a camel train and spent months of hardship scouring the desert surrounding Mount Sinai. Slag dumps thousands of years old contained traces of copper. Many narrow seams of rich oxidized copper ore showed evidences of having been worked countless years ago. But there was no sign of the mineralization on a great scale that he had hoped for. The costly Sinai venture was a failure.

Cyprus was not far away, and Gunther went there next. He was still enthusiastic, though he was beginning to wonder if all the ancient mines were in little scattered veins. After all, the total amount of copper used by the Phoenicians and Greeks combined amounted to very little in terms of modern production. Small deposits might easily have supplied all their needs.

Before going into the field he spent a few days in the old city of Nicosia, trying to get reports or rumors about ancient workings or dumps. Cyprus proved to be much larger than he had thought. It would take months to cover all the ground, and then he might miss the showings he was looking for. By great good fortune it chanced that the Director of the Forest Department of the island was in the capital. He was a keen, intelligent Englishman, and the romance in Gunther's search appealed to him. On his trips about the island he had noticed the large piles of Roman slag, still black and glassy. Unfortunately the Romans had carried the ore to places where charcoal fuel for smelting could be most easily secured. Roman slag dumps did not mean that mines were near by. Other irregular hillocks, covered with grass and trees, might be the dumps of mines, or very ancient dumps of decomposed slag. He helped Gunther to plan an itinerary that would take in these possible ancient workings. Skouriotissa was the first objective.

Gunther started out by carriage over the almost impassable rough roads. It rained dismally, and the food and shelter in the Greek and Turkish villages were even worse than in the Mexican desert. Tired and discouraged, late one afternoon he drove over the last ridge between him and the valley in which lay the almost deserted village of Skouriotissa. As he passed the summit, the setting sun came out from behind the clouds. In its last rays glowed a range of brilliant red and yellow hills — just the colors he had dreamed of. The staining must mean a disseminated ore body. The world was his.

That night he spent in the little native village in the valley at the foot of the hills, the village where, as he happily thought, the miners of thirty centuries ago had lived. Early in the morning he was in the field. The specimens of stained outcrop bore out the hopes of his first distant view, and pellets of copper in the crumbling slag dumps proved that his conclusions were correct. It must be a great "porphyry" ore body. He was soon on his way back to the capital to secure a prospecting permit.

The troubles were not yet over. Long dickering with the colonial government failed to secure a permit that Gunther thought reasonable. There would be no point in spending a great sum in developing the mine if his syndicate could not be sure of a profit in keeping with the risk. He finally gave up hope and continued his explorations without success in Spain and North Africa. Cyprus was the only chance. So he returned to Los Angeles to consult his employers.

Mr. Mudd and Mr. Wiseman had vision enough to enthusiastically approve of the development of the Cyprian prospect and soon won over the other members of the syndicate. They sent Gunther back to Cyprus with authority to start work if he could get his permit. The prospect of immediate development won over the colonial office. The permit was granted, and in 1913 development was finally begun.

Even now Fate refused to grant too easy a success. Gunther's first tunnel broke into Roman workings, evidently some of those from which Galen had described the winning of copper sulphate. Earthenware oil lamps were still in niches in the sides of the drifts. The workings were still open and in fair condition. Only the openings to the surface had been filled with earth and hidden by vegetation in the sixteen centuries during which the mines had been abandoned. The goal had been reached, but there was almost no copper in the rock exposed by the tunnels. It takes very little copper sulphide to make a drip of sulphate-bearing water, and Gunther sadly thought that maybe that was all there had been at Skouriotissa.

The brilliantly stained surface was too attractive to abandon without further development. As tunneling was too slow and the proper level for exploration uncertain, a churn drill was brought over from the United States in 1914. Success was still elusive. The first eight drill holes, in the center of the area that should contain a great disseminated ore body, developed only worthless lean rock with hardly more than traces of copper minerals. The surface staining was caused by slight mineralization with iron sulphide carrying almost no copper. Failure seemed inevitable. As a last resort Gunther decided to put down one more hole through a capping of barren limestone on top of the hill. And this last hole found nearly one hundred feet of solid copper and iron sulphides.

It was an ore body entirely different in character from what he had expected. Instead of having copper and iron sulphides sparsely disseminated through the altered rock, this ore was an almost pure iron sulphide mixed with a little copper sulphide. It could not be treated by the simple concentration process that had brought success to Utah and Ray, and the grade was too low to make direct smelting profitable. The prospect looked black again. But the leaders of the syndicate realized that any deposit carrying 2 per cent copper might be valuable if it were big enough. They authorized further drilling. And the Skouriotissa ore body grew to an amazing size. Within a few months they had developed a lens of solid sulphide 1850 feet long, 800 feet wide and 140 feet thick, averaging 2.1 per cent copper and 48.5 per cent sulphur. A body of this size and grade was one of the world's greatest sources of sulphur as well as a great copper ore body. Costs could be divided between sulphur and copper instead of charged to copper alone.

The resemblance to Arizona climate and geological conditions had proved a will-o'-the-wisp, and Gunther had failed to find the hoped-for porphyry copper deposit. But he had unexpectedly stumbled on an ore body nearly as valuable. It proved to be a rival, not of Utah or Ray, but of Rio Tinto.

The beautiful theory that Gunther started with still looks perfectly reasonable. But his own ore body refused to come under the theory. Perhaps the theory and the girl whose tardiness caused its birth were both instruments of an ironical Fate that uses the most unlikely aids and discards them when they have served their purpose. The girl may have felt that her share in the discovery was even less appreciated than that of the theory, for sad to say, she was not the one whom the discoverer later married.

As underground development followed the drilling at Skouriotissa the mystery of the ancient miners was cleared up. The Phoenicians and Greeks had mined a thin layer of rich "oxide" ore that lay on top of the great sulphide lens and under the beds of clay and limestone that cover the surface. They worked this ore out so thoroughly that hardly traces of it remain. In the succeeding centuries the soft overlying clay gradually sank, filling the workings so completely that only an occasional bit of wood, changed by the ages to charcoal, remains to tell the story. The Romans found ore of a character that they could treat practically exhausted. With their usual resourcefulness they resmelted the Phoenician slags, that still contained a fair amount of copper. As they could not smelt the sulphide ore that their predecessors had left, they let Nature extract the copper for them. They had noticed that water dropping in workings under the worthless sulphide material carried copper sulphate. So they ran a network of tunnels and inclines in this underlying material and caught the copper sulphate solution. As the water level became deeper, they carried their workings farther below the surface, so that there always was a drip they could catch. Sometimes the drifts caved, and crushed or imprisoned the slaves who were forced to do the work. Slaves were cheap, so the dead or dying men were left in the mine, and new inclines were run to take the place of the old. This ingenious method of mining continued until the deepest Roman workings were far down in the barren rock underlying the sulphide.

What ended the ancient mining, no one knows. Perhaps the mines were drained to so deep a level that there was almost no circulation of water through the sulphide, and no copper sulphate formed. Or maybe the Romans continued to work until the decadence of their empire, when their whole civilization collapsed. We only know that for sixteen hundred years Cyprian copper was only a memory.

The rest of the story is the common one of almost hopeless obstacles overcome by engineering skill and financial courage. Millions of dollars were spent in underground development, plant for mining and treating the ore, railroads, harbor building and docks and all the countless details that make up a great industry. Another ore body even richer than that at Skouriotissa was found a few miles away at Mavrovouni. The problem of marketing the iron sulphide by-product in competition with Rio Tinto and other great established companies was solved. In 1922, ten years after Gunther's first visit, the first steamship load of ore was shipped from Morphou Bay. Production rapidly increased until the Cyprus Mines, Inc., was producing several million pounds of copper and twenty to thirty thousand tons of sulphur per month. The courage and patience of the syndicate have been richly rewarded.

Gunther died in Cyprus before the full success had been realized. But he lived long enough to know that his dream in the library had resulted in one of the great copper mines of the world.

AUTHOR'S NOTE: Looking backward from 1973, the story of Cyprus Mines is one of continued success in spite of formidable obstacles. The conflict between the Greek and Turkish peoples on the island often threatened to make operation of the mines impossible. The managers of Cyprus Mines succeeded in convincing both factions that it was to their advantage to allow the mines to keep on running. This was done in spite of the hatred between the races that sometimes made it necessary to change the entire crews of ore trains at the dividing line between areas of Greek and of Turkish influence.

Cyprus Mines has succeeded in finding enough new ore to make a long life assured. At the same time it has succeeded in developing other great mines; the great Pima low-grade copper mine in Arizona is described in the second part of this book. A very large open-pit lead-zinc-silver mine in the wilderness of the Yukon in Canada adds to the record of success. The third generation of brilliant mining executives — first Col. Seeley Mudd, then Harvey and Seeley Jr., and now Henry Mudd — has climaxed a remarkable dynasty.

Col. Seeley W. Mudd, above left, resuscitated the copper mines on the island of Cyprus after they had been abandoned for 2000 years. Below is the Roman slag dump at the Skouriotissa Mine, Cyprus, that attracted the attention of his exploring engineer. *(–All photographs in this chapter, Cyprus Mines Corp.)*

Broken Roman pottery was found near the Skouriotissa mine office. It was used for carrying copper-bearing water from the mine. The ancient Greek theater below, near the Cyprus Mines loading jetty at Morphou Bay, was excavated by Swedish archaeologists about 1940.

Cyprus Mines reopened this Roman tunnel, which had the original timbers still in place. Below is part of a Roman hand winch, or hoist, with a fragment of Roman rope, found in old workings in the Mavrovouni Mine.

The pictures opposite show the Skouriotissa open pit of Cyprus Mines in November, 1967.

The largest Rio Tinto open-pit mine, the Atalaya, shows old underground workings cut by the more recent open-cast diggings. *(–All photographs in this chapter, Leonard U. Salkield)*

2

The Grandeur That Was Rome

THE Spanish peninsula is for the most part a desolate and forbidding land. The beauty of its brush-covered mountains is the beauty of a new country, in which nature has not had time to heal the scars left by too rapid erosion. The struggle for existence in this barren region has always been a hard one. The people have never had a chance to become soft through easy living. Only in a few valleys has Nature relented and built luxuriant gardens to contrast with the austere hills that surround them.

To make up for the general desolation, the Spanish mountains are richly endowed with all the metals used by man. For thousands of years the Iberian peninsula was the treasure vault of the world. Iron, lead, zinc, copper, tin, quicksilver, gold, silver and other rarer metals were abundant. As a result, this desolate land has been rendered even more desolate by a long succession of invasions by peoples that wished to profit from the mines.

The Sevillan plain in southwestern Andalusia is the loveliest and largest of the oases in the waste of tortured mountains that makes up the Iberian peninsula. The rivers that flow southward into the Atlantic have carved down the rugged Sierra Morena and have built out of the mud they carry one of the richest areas in the world. For a hundred and fifty miles west from Seville and the Guadalquivir River to Cape St. Vincente at the southwestern tip of Portugal, and for twenty to sixty miles north from the Atlantic shore to the foot-hills there is a continuous, gently sloping delta of the most fertile soil. The hot summer sun, following abundant rains of the winter and spring, has transformed this whole area into lovely fields and gardens, where groves of figs, olives and oranges are scattered among luxuriant wheat fields. The wide mouths of the rivers have brought the ships of thousands of years to the heart of this plain. The Phoenicians, the Romans, the Vandals, the Moors, and the Castilians in turn built their cities amid the ruins of their conquests. It has always been foreigners who have ruled. Many passing races left their marks on the facial types and on the architecture before the gingerbread house and the motor truck claimed Andalusia for their own.

Nature was not content with the agricultural wealth with which it endowed this corner of Europe. For in the barren hills that form the north border of the plain there are some of the greatest deposits of copper ore in the world. Every new conquering horde rejoiced to find there an abundant supply of the metal it most needed, close to gardens that would feed millions of workmen. For more than three thousand years the border of the Sevillan plain has been one of the great workshops of the world.

Lest any new people might overlook the wealth that lay underground a brilliant trail leads to the copper deposits. The old city of Huelva is nearly in the center of the fertile strip, eighty miles west of Seville. It was from the village of Palos, at the entrance to the harbor of Huelva, that Columbus set sail for America. Flowing into the upper end of the bay there is a little river so remarkable that no one could fail to wonder at it. For the water of the Rio Tinto and the mud that lines its shores are a brilliant brick red. For sixty miles this strange red stream wanders through the wheat fields and orange groves. Then, just as it emerges from a shallow cañon in the foot-hills, the water and the mud suddenly change to green! And just above the point where the change takes place are the great ore bodies of Rio Tinto.

A great deal of the color is due to impure iron-bearing solutions that have percolated through the great leach-heaps placed on the banks of the Rio Tinto by modern miners. But natural leaching must have made this river, with its brightly colored banks and bitter, poisonous water, a source of curiosity and perhaps of fear to the Phoenician explorers of three thousand years ago. Curiosity evidently overcame the fear, and the adventurers who sought the cause of the wonder were rewarded by the greatest mines of ancient times.

Mining in the Rio Tinto district started far before the dawn of history. The earliest inhabitants, of whom only a little is known, were a savage race called the Iberians. Even this primitive people had gold and silver ornaments. The precious metals had been concentrated by weathering on the surface over the ore bodies and in the nearby streams. All the first miners had to do was to pick up the bright metallic pebbles that nature had collected for them.

The Iberians offered little resistance to the expansion of the Phoenician traders who followed them. Nearly four thousand years ago, ships from Tyre were sailing between the Pillars of Hercules in search of new markets and new sources of raw materials. Among their many colonies was Tartessus, where Huelva now stands. As early as 1240 B.C. the enterprising Phoenicians had started to mine the great Rio Tinto deposits. Little is known of their operations, and only small heaps of reddish decomposed slag remain as evidences of their work. Very likely their first efforts were devoted to winning gold and silver from the capping of porous iron oxide that overlay the copper ore bodies. Long ages of weathering had dissolved the copper and sulphur from this upper shell, leaving the precious metals in a concentrated form. Gold and sometimes silver could easily be won from this soft decomposed rock by even the most primitive methods of hand washing. A few feet deeper below the surface there were rich oxidized copper ores. The Phoenicians could mine these, too, in quarries, recovering the copper in pure form by melting the ore in small charcoal furnaces. The industry grew so rapidly that in 1100 B.C.

Rio Tinto copper was one of the important resources of the Phoenician Empire. The merchants and navigators of this great seafaring race brought tin from the mines in Cornwall, melted it with copper from Rio Tinto to make bronze, and traded the bronze for the choicest products of all the peoples around the Mediterranean.

The Phoenicians were content to leave their colonies alone as long as trade was not interfered with. Instead of annihilating the Iberian natives in southern Spain, they peacefully assimilated them. With the luxurious customs of Tyre emplanted in this rich and sunny land, life in Andalusia must have been pleasant twenty-five hundred years ago. But the Phoenician Empire gave birth to one far harsher than itself. Carthage, first a Tyrian colony, grew in a few years to dominate all the ancient world save for the new and struggling Roman Republic. Under the Carthaginian rulers, Spain was rapidly built up into a great industrial colony. Hamilcar introduced systematic mining by great armies of slaves in place of the simpler, less arduous methods of the preceding centuries. Rio Tinto was busy, and a few aristocrats were prosperous, but life was hard for the workers in the fast deepening mines.

As the power of Carthage waned before the assaults of Rome, the first period of prosperity at Rio Tinto came to an end. Phoenician ships could no longer sail the seas unhindered. With shipping that carried the copper to the markets of the world destroyed, the mines were abandoned. Rio Tinto became the first of the deserted "ghost cities" that are the sad monuments to worked out mines the world over.

In 210 B.C. Publius Scipio Africanus finally drove the Carthaginians from southern Spain. Then followed six hundred years of the most unbroken peace and prosperity that any country has enjoyed. The soldiers of the Legions sent to pacify the country remained as colonists, often marrying the native women. Before many years Spain had become completely Romanized, and its civilization vied with that of the capital itself. Even the civil wars that followed Caesar's death made little impression on this happiest part of the empire. At first the rich agricultural valleys were the greatest asset of Hispania, and large exports of grain were sent to Rome. But with the growth of the empire and the rapid expansion of the industrial civilization that accompanied it, the Spanish mines rapidly came into prominence. There was an almost unlimited demand in the great Roman cities for all of the metals, of which the Iberian deposits were the most abundant source. Roman genius for organization and for engineering soon surpassed the achievements of the Carthaginian industrial age. The conquest of other parts of the growing empire sent millions of slaves to labor at the mines and furnaces. They built great cities to house the workmen and the armies that guarded them. Arching aqueducts that are still in use brought water from distant mountains. Every foot of the rich valley was intensely cultivated. Soon the population far exceeded the capacity of the country to produce food. At the height of the Roman power, in the second and third centuries A.D., there were nearly fifty million people in Hispania. Now hardly thirty million win a bare subsistence from an overcrowded land.

To supply this busy multitude, a thriving commerce grew up. For hundreds of years the Mediterranean was alive with ships carrying wheat from the rich fields of Tunis and Algeria to Spain, metal wares from Spain to Rome and the other cities of the empire, and slaves and colonists to the African agricultural centers. The millionaires who owned the ships and the mines lived in luxury in the capital, firm in the conviction that so great an industry and so powerful an empire could never decline.

Modern minds recoil at the terrible lot of the slaves who built up the greatest of colonies. We think of them crawling away to die like vermin in the dark passages in poorly ventilated mines. The long story of massacres, gladiatorial fights, and assassinations that mars the history of the Roman Empire proves that no pity could have softened the lot of the conquered peoples who were dragged far from their homes to labor for the glory and wealth of the patrician families. But as the years passed, even the lives of slaves became less miserable. Economic necessity supplied the place of the altruism that had not yet emerged from a world of war and rapine. The empire ceased growing, and as a result slaves became scarce. When they were killed it was expensive to replace them. To prevent the exhaustion of these necessary cogs in the production of Roman wealth, labor laws were enacted and strictly enforced.

Bronze tablets inscribed with the mining regulations of the Romans have been found at Aljustrel, just west of the Portuguese border. They provide a complicated system of royalties to the state. The lessees were required to keep the workings well timbered and safe and to leave pillars around the main passageways. The slaves were well treated, as only the governor could order them whipped. Their sleeping accommodations, food, hours of work, and even facilities for bathing were ordered in great detail. The worst punishment was to sell a slave with the provision that he could never again work in the mines. Mining must have been popular in those days. For all our boasted liberty and regard for the rights of others, these Roman laws for the treatment of slaves are more humane and liberal than the Spanish labor laws of the present day.

Of all the centers of Roman operations in Spain, Rio Tinto was by far the greatest. The Carthaginians had only scratched the surface. Stringers of gold and silver ore still remained in the bright red capping of iron oxide that covered great barren areas in the low hills. A little deeper below the surface, far larger bodies of rich oxidized copper ore were found, forming an irregular blanket between the surface iron oxide and the enormous lenses of sulphides that have been the seat of modern operations. Still deeper, bands of the rich oxidized ore ran far down into the sulphide lenses, following fractures or cracks that permitted the surface water to penetrate more deeply than usual. These ore bodies were soon the chief source of copper in an empire that depended largely on copper and bronze for its supremacy in the arts of war and peace.

In every branch of the mining industry the Romans made rapid strides. In the first essential — the discovery of the ore —

they even surpassed modern science. How they did it, we have no idea. Their skill was so taken for granted that the old chronicles simply state that "the development of the ore bodies is carried on in the usual manner, and presents no difficulty." At any rate the difficulty was not insurmountable, for in the copper district that stretches from Seville to beyond the Portuguese border the Romans found nearly a hundred lenses of ore. All the skill that modern engineers can muster has added only three or four comparatively small bodies to the number, and the great production of the past fifty years has come almost entirely from ore bodies that the Romans discovered.

With cheap slave labor, the natural assumption is that the Romans sank prospect shafts indiscriminately so close together that they could not miss the ore. In the heavily mineralized areas where the brilliant iron oxide capping of the ore bodies came to the surface, this was actually done. Any shaft would find low-grade material, and the Romans honeycombed the red hills in order to find the richer streaks. Only a little observation and common sense were needed to outline the areas that might contain ore. But in the southwestern part of the mining area the ore zone extends out under the Sevillan plain. There the outcrops of the ore bodies are effectively hidden by several feet of soil and then by ten or fifteen feet of a barren layer of limestone that was laid down long after the ore was formed. No sign of the underlying minerals could penetrate this mask of limestone and earth. Yet here, too, the Romans sank their pits straight down through the soil and limestone into the mineralized formation. Hardly a Roman pit can be found that did not enter ore or low-grade material that was worth prospecting.

Even with the recent aid of magnetic and electrical prospecting, the engineers of the present boasted age of science have not succeeded in finding a single new ore body under the Sevillan plain. The Roman geologists knew something that our research has not been able to recapture.

It is interesting to speculate on the secret of the ancient discovery of ore bodies, even at the risk of becoming a geological apostate. For the speculation leads beyond the accepted realm of science into the scorned field of the quack and the charlatan. Can it be that there is something in nature even more delicate than the waves the radio receiver of the galvanometer can detect? The suggestion must be made half in jest, but with the uncomfortable feeling that maybe the joke is on the present-day geologists. A successful Spanish engineer with an unusually keen imagination has come upon a possible explanation of the Roman prospecting method that will at least convince our Doctors of Philosophy of the abysmal depths of Roman superstition and ignorance. He noticed that in all the ancient paintings two of the earliest Christian saints in Spain, St. Abdon and St. Senen, hold in their hands twigs that look very much like the willow wands of the "water witch" to whose prowess so many of our own country people trust in locating wells. Can it be that the Romans also had faith in the strange art of dowsing, and that the dowsers really helped them find the ore bodies? If cattle in the desert will infallibly scrape away several feet of sand in a dry wash to find the place where a hidden stream of water comes within reach of the surface, may there not be some emanation of energy from ore bodies that certain persons, with a sense that is undeveloped in most of us, might discern? The willow wand may simply be a means of making effective this subconscious power, like automatic writing or other half-understood mental phenomena. We hesitate to cast aside our disdain of such an unscientific procedure. And yet recent delicate instruments have proved that electrical currents are generated within ore bodies. The mind should be as sensitive as the galvanometers it plans. Who knows but that the skepticism that comes from a little learning has blinded us to the more obscure uses of the mind, and that the radio ore-finder is a clumsy way of imitating the far more effective "nose for ore" with which nature has endowed us. At any rate, the Romans seem to have agreed with our modern miners in calling the one who finds an ore body a saint and the one who fails a rank imposter.

While the Romans took the palm in prospecting, modern engineers can claim superiority in the art of mining. This is largely because we have the benefit of the Chinese discovery of gunpowder. Breaking rock was an arduous task without explosives. Wherever the Romans could find soft seams, they picked ahead of the working faces and wedged or hammered the solid rock into the openings thus made. Quicklime, forced into cracks and then wet, expanded and broke off fragments of rocks. Where the material was exceedingly hard, with no soft seams or cracks, the Romans built fires against the faces, and then threw water on the red-hot rock. This splintered the densest material. It was a slow process, but with abundant slave labor speed was not essential.

The lack of power was as great a handicap as lack of explosives. In removing the broken ore from the mines, the Romans used much the same methods that are still employed in remote districts in Mexico or in South America. They cut steep spiral stairways in the rock round the sides of the shafts or of the stopes – the cavities from which the ore had been removed. Notched poles served for ladders where there was no room for stairs. Oil lamps from their niches in the rock gave a flickering half-light. Panting and coughing in the ruddy smoke, the slaves packed up on their backs leather buckets filled with two hundred pounds of ore. At the end of a thousand-foot climb their lungs were bursting and their breath came in tortured groans that echoed through the dark caverns of the mine like cries from Hell. We shudder at the inhumanity of it. Yet today we can see the same process in remote Spanish-American mining camps. Where wages are a few cents a day, men's lives are often cheaper than fuel for hoists. The Romans at any rate had no alternative. Later they learned to use the hand-turned windlass, and hoisted the ore up vertical shafts in bronze-rimmed leather buckets. Modern workings still occasionally uncover the rims of these ancient buckets. The amount of ore hoisted per man by the Roman methods was trifling. But the multitude of slaves brought out a tonnage that would make a big mine even today.

Underground water was one of the most serious problems of the Roman miners. As long as the flow was small, slaves packed the water out in buckets. In the wetter mines even a steady stream of slaves could not keep the workings dry. Then they excavated chambers and installed oaken water wheels. These were about twenty feet in diameter. Around their rims were wood or leather buckets placed as close together as possible. Cleats were attached to the outside of a wheel, to give the slaves a chance to use their full strength and weight in turning it. As the wheel turned, the buckets dipped up water from the lower part of the chamber, lifted it to the top of the wheel, and poured it into troughs that carried it away. In places the wheels were placed in series, one above the other, so that water was lifted sixteen feet at a time for considerable heights. This is the earliest known form of pumping machinery. Wheels that are still used for lifting water for irrigation along the Nile are almost exactly like the Roman wheels found at Rio Tinto.

The Rio Tinto Company some years ago reconstructed a pumping wheel, exactly duplicating a Roman one that was found in good condition. It is a strong, capable looking machine that should handle a very respectable flow of water. Attempts to measure the capacity by an actual test failed completely. For the loaded buckets were so heavy that the modern workmen, crowded together as closely as possible, could not turn the wheel. It remains a mystery how the Romans turned it. There was no room in the pump chamber for any apparatus that would enable more men to work — such as a bar working over the cleats — and there was room for only a limited number of men about the wheel itself. Maybe the barbarian slaves were more powerful than modern miners, or the threat of the lash was an effective spur.

At last the Roman workings at Rio Tinto became too deep to be unwatered by the combined effect of wheels and water-carriers. A long drainage tunnel was the next device to keep the mines open. It was over a mile long and at the end a thousand feet below the surface. Much of the tunnel was in

This 2000-year-old Roman water wheel was found in the North Lode of Rio Tinto workings in June, 1886.

soft rock, that has since caved and solidly filled the opening. But the ancient workings that have been rediscovered prove that the Roman surveyors made as good a connection with their shaft workings as we could make with the help of transit and steel tape. In underground surveying as in planning aqueducts the Roman engineers could learn nothing from us.

In the fourteen hundred years since the Romans abandoned Rio Tinto, time has obliterated most of their underground workings. Only the shafts that were well timbered with hewn oak are still in fair condition due to the preserving effect of copper and iron sulphate. The untimbered openings from which the ore was removed have long since been filled by falling rock. The millions of tons of slag from ancient furnaces are conclusive evidence that the ancient mines were great ones. There are Roman drifts here and there throughout the whole area of the modern Rio Tinto ore bodies, covering a length of two miles and a width of one mile, and the deepest known workings of the ancients were more than a thousand feet below the surface. Even in our day, the Roman operations rank with the greatest copper mines.

It seems to us incredible that mines a thousand feet deep, that produced many million tons of ore, were worked without dynamite and without power. For power and explosives have revolutionized the art of mining. Where the Romans laboriously wedged off fragments of rock, or splintered an inch or two ahead of a face by alternate fire and water, dynamite will blow out the ground six feet ahead of a small drift, or shatter hundreds of thousands of tons of solid rock in a quarry in a few seconds. And where a slave painfully strained up a thousand feet of ladder or winding stairway in half an hour carrying two hundred pounds of ore on his back, a modern hoist will lift five or six tons in half a minute. Power and explosives have increased the speed of mining a hundredfold. What we forget is the fact that with cheap labor, they have not greatly decreased the cost. A Roman slave cost only the labor of another slave to feed him, which was of no importance as long as slaves were plentiful. Even in Roman times Rio Tinto copper must have been cheap copper.

In the final stage of the mining industry — the recovery of the metal from the ore — the Romans equaled modern science in everything save size of smelting furnaces and the treatment of sulphide ore. Except for a little copper recovered from sulphate solutions derived from the sulphides, practically all of their production was from the upper part of the ore bodies, where the sulphur that is combined with the metals in the deeper, unaltered ore has been slowly burned off by the atmospheric oxygen in rain water that seeps down from the surface, leaving oxides of iron and copper in place of the original sulphides. As the resulting "oxidized ore" occurred nearer the surface, in ample quantities, the inability to smelt sulphides was not a serious handicap. The small smelting units were a nuisance, but they were remarkably efficient. In place of our great smelters that treat fifteen hundred tons of ore a day in one furnace with every stage of the operation mechanical, they had a multitude of small furnaces operated by hand, smelting at most a few hundred pounds of ore at a charge. The slag from these furnaces covers hundreds of acres on the hills around the Rio Tinto. Engineers estimate that there are from twenty to thirty million tons of it. The Roman smelting was so efficient that the slag often contains hardly a trace of copper, and seldom contains as much as the slags made by the best modern copper furnaces. A few good Roman smelter men would be in great demand at a time like the present, when all the copper companies are using every effort to increase efficiency.

The immense tonnage of Roman slag at Rio Tinto and its remarkably low copper content have never been satisfactorily explained. Millions of tons of charcoal were required to smelt the ore. Yet the country surrounding Rio Tinto is covered only by low brush. Where could the charcoal have come from? Part of the barrenness is due to the effect of sulphur-burning fumes from roast heaps of modern times, but the untimbered area is too great to admit of this as the full explanation. Were great forests completely destroyed by the ancient charcoal burners, or was charcoal packed in from forests many miles away? Either answer seems unlikely. The very low copper content of the slag is just as hard to explain. Oxidized ore usually results in a slag that contains a lot of copper. Therefore some authorities think that some ore other than copper ore was smelted at Rio Tinto. This suggestion is borne out, they say, by the fact that the ancient miners could not possibly have produced twenty or thirty million tons of oxidized copper ore, and ancient industry could not have absorbed the one to five million tons of resulting copper. They think the Romans may have brought lead and zinc ore by boat and wagon from Cartagena, on the Mediterranean two hundred miles east of Huelva, and smelted it with the abundant supply of charcoal that they assume Rio Tinto once enjoyed. Or possibly they smelted a large amount of the iron capping above the copper ore for its gold and silver content.

Some day a search of ancient manuscripts may throw light on the way the Romans found their ore and on the source of the ancient Rio Tinto slag dumps. Until then we can only wonder, with deep respect for the people who found and efficiently worked some of the greatest copper mines of the world without explosives and without power.

Spain was remarkably free from the wars and insurrections that filled the last two hundred years of the rest of the Roman Empire. There was much grafting and oppression by the political rulers and industrial magnates, but the centuries of peace more than atoned for these minor troubles. As always, peace and prosperity brought with them a high level of civilization. All of the arts flourished, and many of the leading writers and statesmen of the time were born in the Spanish colonies. Seneca and Martial added to the glory of the whole Roman Empire as well as to that of their native cities in Hispania. Even the more intelligent slaves shared in this pleasant culture. The descendants of the Gauls who were brought in to labor in the mines and orchards became tutors in the patrician families. No other part of the empire presented such untroubled opportunities for advancement.

Yet for all this enjoyment of the happy arts of peace some economists think that Spain was in a large degree responsible for the fall of Rome. Wealth from the Spanish mines was needed to pay the mercenaries who guarded the empire, and to keep the industrial civilization of the capital running. The mines seemed for a time inexhaustible. But gradually the richer ores were mined out. The gold and silver production rapidly fell. The remaining copper ore was largely sulphide, and the amount of copper sulphate recovered from water that had seeped slowly down through this ore was far too small to meet the demand. The dwindling supply of metals added financial panic and hunger to the troubles of the dying empire. The forges of the Roman Workshop one by one grew cold. By the time the Vandals completed their conquest of

Spain, in the fifth century, Rio Tinto had once more become a ghost city.

Of the three centuries during which Spain was ruled by the Vandals and Visigoths, Rio Tinto has left no record. The Roman fortresses crumbled into ruin. There was probably a small production of copper from the sulphate-bearing waters, just as a few hopeful leasers linger on in the exhausted modern mining camps and make a bare living long after the great mines are dead. But the glory of the Roman days had departed.

During the Middle Ages, the Spanish mines continued their strange record of always serving foreign masters. This time the Moors were the conquerors. Their beautiful buildings are the most striking features of southern Spain. In iron and steel working they excelled all the people of their time. They used copper largely for cooking utensils and for bronze ornaments. Therefore it was of less importance to the Moors than to their predecessors. But in their five-hundred-year rule in southwestern Spain they made important advances in the treatment of copper ores. Oxidized ores had been about exhausted before the Moors came. As they could not find any more ore bodies they needed some way of recovering copper from the sulphide ore much more efficient and rapid than the Roman method of collecting sulphate solutions that slowly seeped down through the ore bodies. This they succeeded in doing. They found that if they mined the sulphide ore and placed it in heaps, and allowed water to percolate down through the heaps, this water contained far more copper than the underground water. The Moorish scientists, with their skill in mathematics and all the theoretical sciences, may have recognized that exposure of the broken ore to the air was responsible for this. The copper sulphide must be oxidized to sulphate before water will dissolve it, and a free circulation of air through the broken ore causes this oxidation to take place rapidly. The Moors also found out that if the water containing copper sulphate is allowed to run over iron, pure copper is precipitated, and the iron is dissolved. As iron was cheap and abundant in Moorish Spain, this discovery completed an efficient method of recovering copper from sulphide ore. The process is so well adapted to Rio Tinto ores that it is used with almost no changes today.

With the recovery of the metal from sulphide ore perfected, the Moors had all the copper they needed. The Rio Tinto district was prosperous again for hundreds of years. Operations were on a far smaller scale than during the Roman occupation, because the market for copper and bronze was comparatively limited. The mining towns were less crowded, and the adjoining plain could easily feed the miners. Great bodies of the sulphide ore came close to the surface. The deep, poorly ventilated Roman shafts, following oxidized seams, were no longer needed. Instead the miners could work in open quarries under the blue Andalusian sky, and run the ore out on the leaching heaps without painfully carrying it up the steep ladders and winding rock stairs that had broken the hearts of so many Roman slaves. More attention could be paid to living now that there was no longer the urgent Roman demand for efficiency and great production. The Moorish period must have been the happiest in all the long story of Rio Tinto.

After Ferdinand III of Spain drove the Moors from Seville in 1248, Rio Tinto fell on evil days once more. For two hundred and fifty years there was still the menace of Moorish raids as the Saracen capital, Granada, was less than two hundred miles away. The Spaniards were too busy with constant wars to carry on important mining operations. Even after Ferdinand and Isabella drove the last Moorish invaders from Spain in 1492, there were more exciting adventures than running copper mines at home. Why dig copper when gold and silver beyond the wildest dreams were waiting to be picked up by any who ventured to cross the Atlantic? Only a few of the less ambitious miners stayed home and kept the leach heaps going in a desultory way. Before long the Moorish mining camps decayed into picturesque ruins.

Even under the Spanish rule it was foreigners who finally brought another era of prosperity to Rio Tinto. The Spaniards were intent on developing the richer American mines. But adventurers from northern Europe found that in spite of all the centuries of mining the ore bodies in Spain were still richer than those in their own countries. The Spanish crown was glad to give leases on the idle mines, in the hope of finding new industries to tax. In the seventeenth century a Swede named Wolters got possession of the principal properties and produced copper with fair success for many years. He was succeeded by Ticquet, a Frenchman. In the eighteenth century Lady Maria Theresa Herbert and a company of English adventurers secured the lease, and worked the mines profitably until the French invasion of 1800 put an end to the operations. In 1812 the government tried to start mining again. It let a succession of leases that yielded insignificant production or profit. The Age of Electricity had not yet begun, and the world had little use for copper.

The final period in the long story of Rio Tinto began in 1873. Then the London firm of Matheson and Company bought the mines for 92,800,000 pesetas, then worth about £3,712,000. They organized an English corporation called the Rio Tinto Mines, Ltd. This company has worked with increasing success up to the present.

A new development in the method of recovering copper from the sulphide ore was largely responsible for the success of the Rio Tinto company, as well as of many other companies in the great copper district of southwestern Spain. The heaps of sulphide ore oxidized very slowly. The earlier miners had to leach the heaps for several years before they made a fair recovery of the copper. Therefore the cost of mining had to be paid several years before they could sell the resulting metal. The investment in broken ore was too great to allow much profit. There seemed to be no way to avoid this expense. But one day a pile of brush was burning at the edge of a sulphide heap. To the surprise of the workmen, the ore itself caught fire and burned vigorously for several weeks. That ore was spoiled, they thought. But they turned the water on to see if they could still recover any copper. The water came out at the toe of the heap far richer in copper than any solutions they had seen before. They soon learned that if they burned, or roasted the ore before leaching, they could recover more copper in six months than they used to get in three years. They had replaced the slow oxidation of natural rusting or corrosion by the rapid oxidation of burning. With the investment in broken ore cut by more than 80 per cent, Rio Tinto became a low-cost copper mine.

This plan of burning or roasting the sulphide heaps to hasten the oxidation made southwestern Spain again, for a brief period, the greatest copper district in the world. The demand for copper throughout Europe was rapidly growing with the advent of the Industrial Age. Here was an almost unlimited supply of copper right at the door of the great markets, occurring near the surface in enormous ore bodies that could be easily mined in open quarries. And the simplest process ever developed would recover the copper cheaply and quickly. No wonder the Rio Tinto and neighboring companies prospered. Heaps of millions and then tens of millions of tons of ore covered all the available space on the hillsides near the mines. Sulphur dioxide gas from the burning pyrite filled the air with strangling vapors that set men and animals coughing and killed off every green thing it touched. And the green sulphate solutions flowed in ever-increasing quantities from the foot of the heaps, depositing bright crusts of pure copper on the iron in the precipitating troughs and adding to the red burden of iron oxide that gave the Rio Tinto its name.

Luckily the prevailing wind was from the coastal plain toward the mountains, so the sulphur smoke did not ruin the rich fields and orchards. But the mountains and foothills north of the mines were not so fortunate. The clouds of white vapor killed every tree within ten or fifteen miles. Life was no longer pleasant as in the time of the Moors. Driven by their employers' lust for more copper and more dividends, the workmen toiled for eleven or twelve hours a day in return for a few cents, coughing their lungs out in the midst of a desolate waste where no plant could grow. It was the Industrial Age at its ugliest.

Since the beginning of the present century, a new development in the treatment of the ores is fast redeeming the Huelva area from its reputation as the harshest of mining districts. When the ore was roasted, many million tons of sulphur went off in the air from the burning heaps as sulphur dioxide gas. But sulphur is one of the most valuable products of modern industry. From it is made sulphuric acid, the basis of most chemical products and of the phosphate fertilizers that keep the time-worn European fields from becoming barren. The acid factories used to burn the pure sulphur that occurs around old volcanoes, especially in Sicily. But burning iron sulphide, or pyrite, makes the same gas as burning sulphur. With no additional mining cost, the value of Rio Tinto ore was doubled, and the sulphur that had ruined the Roman mines became the greatest asset of southern Spain.

The mines were lucky in winning the additional profit at almost no expense. For leaching in heaps does not hurt the pyrite that contains the sulphur. The miners of the present day dig out the ore and place it on the hillsides in heaps of a few hundred thousand tons. By alternately leaching the ore and letting it dry and oxidize, they recover half the copper in a year, and 90 per cent in three years. With better control of leaching they recover the copper much more quickly than was possible a few years ago. As soon as the amount of copper remaining in the ore becomes too small to be profitably recovered, they load the remaining pyrite into railway cars with steam shovels, haul it to Huelva, and there load it in ships that carry it to sulphuric acid plants all over the world. Some of the mines smelt the richer copper ore directly, without saving the sulphur. But the thousand-year-old heap leaching process, followed by shipment of the residue for its sulphur content, has made the Rio Tinto area for the fifth time one of the greatest mining districts of the world.

It was time some new asset was found for the old mines, for the better copper ore occurred in a layer only two or three hundred feet thick. Under it the iron sulphide continued to far greater depth, but the percentage of copper fell from

3 or 4 per cent, to less than one per cent. The richer ore is almost exhausted, and the few pounds of copper in the underlying lean material will not pay the cost of mining. The discovery that acid can be made from the iron sulphide saved Rio Tinto from gradually declining until it once more became a ghost city.

Even the Romans would be amazed at the changes at Rio Tinto in the past thirty years. At the seven great sulphide lenses there are now yawning quarries that have cut deep gashes across the hills and valleys. One of these cuts is two thousand feet long, nine hundred feet wide and three hundred feet deep. Steam shovels that would seem to the Romans like grotesque, powerful gods tear up several tons of ore at a bite and drop it into railroad cars or into raises that lead down to the mine haulage levels. Every year a few hundred workmen mine and treat as much ore as the crowded thousands of Roman slaves laboriously produced in fifty years.

And the end of Rio Tinto is not yet in sight. Even at the madly accelerated modern rate of production, the Roman Workshop will be a great center of industry for many generations. The Phoenicians mined a few hundred thousand tons of ore in a thousand years. The proud Romans dwarfed the efforts of their predecessors by mining twenty or thirty million tons in half the time. And now the moderns have taken from the same ore bodies a hundred and fifty million tons in less than fifty years. In prosperous years Rio Tinto alone produces two and a half million tons of ore. Six of the great lenses were apparently exhausted a few years ago, and the end of the old district seemed near. But the seventh lens, called the Atalaya, has surprised everyone by greatly expanding below a depth of seven hundred feet, with some of the richest ore Rio Tinto has known. And the modern flotation process has made valuable several million more tons of lean material around the sulphide lenses that was thought to be worthless a few years ago. In spite of the enormous production of the past two thousand years, the Rio Tinto mines alone still have two hundred million tons of developed ore — far more than their whole production up to the present.

Even the gargantuan scale of modern operations has not obliterated the evidences of Roman skill. The past and present are brought into striking contrast by one of the strangest sights in the world — the relentless teeth of a steam shovel tearing out the oaken timbers of a Roman shaft.

AUTHOR'S NOTE: In the four decades since the chapter on Rio Tinto was written, this remarkable operation has been forced to overcome political as well as technical problems. The settlements with the Spanish government and the changes in production methods are outlined in the second part of this book.

The white sulphur dioxide fumes from burning leach dumps at Rio Tinto in 1900 indicate that serious air pollution existed many years ago.

On the page opposite is the open pit at Rio Tinto about seventy years ago. Below are a Bucyrus steam shovel and ore train with steam engine of approximately the same time. The shovel, with a 1¾-cubic-yard bucket and weighing 95 tons, was one of the largest in use early in the present century. It would be dwarfed by modern machines.

Day shift at the Phoenix Mine with an ore train was photographed in the early days in the Copper Country of Michigan. *(–All photographs in this chapter, Michigan Technological University Library)*

3

The Copper Country

WHEN the first white men came to America, they found the Indians in the depths of the Stone Age. Arrow- and spear-heads were made of carefully chipped flint. Needles and knives were shaped from bone, and dishes were made of wood or pottery or basketware. In the midst of this primitive culture, the early explorers were surprised to find fairly abundant metal objects. Henry Cabot reported that the natives along the American coast had "plenty of copper as beadstones hanging at their ears." Henry Hudson in 1609 saw tobacco pipes and necklaces of red copper. A few of the tribes even had metal axes and knives. These tools and ornaments were not made of brass or bronze, like the relics of earlier civilizations in other continents. They were all hammered out of remarkably pure copper.

The early settlers were so busy fighting the Indians that they paid little attention to the possible source of the copper. The early Jesuit priests and explorers were the first to note that most of it came from the south shore of Lake Superior. English explorers and adventurers followed the trail a hundred years later, but lost it within a few feet of success. Almost another century elapsed before the ancient copper mines were at last discovered, in the 1840s. Within a decade after the discovery the northern peninsula of Michigan became the Copper Country — for a generation the most prolific source of copper in the world.

Geologists have found an interesting reason for the long delay in solving the problem of the source of the Indian copper. The great North American glacier was at the bottom of the difficulty. All school children know that a few thousand years ago a capping of solid ice hundreds of feet thick covered all the eastern portion of the continent as far south as the Ohio River. This sea of ice moved slowly and irresistibly, generally toward the southeast, planing off the hills and mountains and carrying with it billions of tons of partly ground up rock and gravel – the shavings from the glacial erosion. When the ice cap finally melted, it left most of the area it had occupied covered with this ice-formed gravel, which is called by geologists the "glacial drift." Only here and there remnants of the rock that originally formed the surface, rounded and scratched by the moving ice, emerge from this gravel mantle.

Some of the boulders in the glacial drift were carried hundreds of miles before the ice melted and dropped them in their present positions. Therefore the fact that a boulder of a certain kind of rock is found in the gravel at a certain place does not mean that the rock originally came from there. It may have been carried by the ice five miles, or five hundred miles. If the boulder is an ordinary rock, it makes no difference where it came from. But if it is rich ore, it becomes very important to trace it to its source.

This is one of the most intricate detective problems that has been presented to the imagination of geologists. Sometimes the solution has proved impossible. Fragments of the ore may be so widely separated that they can not be found in great enough numbers to show the direction of flow. Or there may have been cross currents in the ice stream, that have left the boulders in a hopeless jumble. More often, when the possible prize has been worth the study the problem has been triumphantly solved. The geologist finds a few boulders of ore strung out in a more or less definite direction. He looks more carefully along the projected course of this line, digging trenches where the surface is covered by soil or vegetation. If his line is the right one, he is rewarded by more and more chunks of ore. The scent is becoming warmer. Then — perhaps after a trail a hundred miles long — it is lost completely. Careful trenching fails to find another fragment of the ore. This means that the end of the hunt is near. The source of the boulders of ore must be so close by that all the fragments torn from it by the ice were carried away before the glacier melted. To find the ore in place in solid rock the geologist carefully examines the nearby ice-polished outcrops that rise above the mantle of glacial drift. If this fails he puts down shafts through the gravel to prospect the area where the rock is hidden. If the glacial drift is too deep or too heavily soaked with underground water to permit shaft-sinking, drill holes through the gravel are the next resort. The hunt may still be unsuccessful, as the nearest fragments may have been carried miles from its source. But if the mass of ore is a large one, and if the geologist is lucky as well as skillful, one of the shafts or drill holes finds the ore body whose ground-off top was responsible for the long trail of fragments in the glacial gravel.

The origin of the Indian copper was one of the most interesting of these problems in glacial geology. Copper articles were most common among tribes that lived in what is now Michigan, Wisconsin and the north shore of Lake Superior. The much scarcer articles owned by other tribes might easily have been acquired by capture or trading. It looked as though there might have been several copper deposits, scattered over a length of three or four hundred miles. But then someone noticed that the homes of the Indians who possessed the most copper lay in a general northwest-southeast line — the average direction of the flow of the old ice cap. And copper became much more common toward the northwest as far as the south shore of Lake Superior, then scarcer again on the north shore. In the Keweenaw Peninsula of northern Michigan copper articles were very abundant. Finally pits were found that had existed long before the white men came. In the bottom of some of them, still in the glacial gravel, there were large boulders of pure copper. In other pits the copper was in place, in chunks in the solid rock.

The mystery of the copper used by the Indians was solved. The original deposits were veins in the Keweenaw Peninsula, on the south shore of Lake Superior. Here pure copper in large masses came to the surface. During the glacial period, a great thickness of rock was eroded away by the slowly moving ice, which tore off the copper as well as the enclosing rock. The bits of copper with the rest of the ground-up rock were carried southeastward, some of them for hundreds of miles, before the ice melted and deposited them in the glacial drift. Thousands of these copper pebbles were exposed in the gravel beds of streams. The Indians, attracted by the weight and beautiful red color of these strange stones, picked them up, and slowly learned how to beat them into the desired shapes. They may have even realized that the hammering made the copper harder. But they never learned the more advanced arts of melting copper and casting it into the form they wanted, or of combining copper with other metals to form harder alloys.

Far from the source, near the Ohio River, there were comparatively few copper boulders. There the Indians had only a little of the metal, which was used almost exclusively for ornaments. Further northwest the boulders became much more frequent, and many copper articles were used. In the Keweenaw Peninsula itself there must at one time have been a flourishing prehistoric copper industry, as copper was present in almost unlimited quantities.

No one knows what race of Indians, if they were Indians, carried on this copper mining. Archaeologists have guessed that they might have been Norsemen, or Mayas, or Aztecs, instead of Indians. Whoever they were, they must have had a fair degree of civilization. They left well-shaped copper tools as well as stone hammers in the ancient pits. For the most part they worked boulders of copper in the glacial drift near the vein outcrops. But they also mined the copper in the upper part of the veins themselves, prying out the masses of convenient size and leaving those that were too heavy to work. They became such skillful prospectors that they found the outcrops of most of the great ore bodies in the district. Some of the ancient workings followed the veins for twenty to thirty feet below the surface. In one pit, sunk in the outcrop of what is now the Minnesota Mine, the primitive miners had lifted a mass of copper weighing six tons on a bulkhead of crisscross timbers several feet above its original position and then abandoned it — perhaps because they could not cut off the copper from the large mass without excessive labor.

The Lake Superior copper was perfectly adapted to use by a primitive people. It was pure copper, and not a chemical combination with sulphur or carbonic acid or silica, as in most copper districts. So it did not require the accidental discovery and slow development of smelting that delayed the advent of the Copper and Bronze Age in many parts of the world. The pebbles were in convenient sizes for all the early uses. Copper of such unusual purity was malleable. A little hammering was all that was needed to work it into the desired shapes. This hammering made the copper much harder, tempering it as a blacksmith tempers steel, but without the heat treatment that a blacksmith uses. When tempered in this way it made knives much better than any of the bone or flint knives the Indians had used before. With the copper occasionally came equally pure silver — not alloyed, but with one end of a pebble copper and the other silver. These bits of bright red copper and shining white silver made beautiful ornaments. The people who had the attractive copper and silver beads as well as metal knives and axes must have felt a great superiority over the neighboring tribes that were not so fortunate.

With a metal of such beauty and usefulness available on the surface in almost unlimited quantities, it would seem natural for the Indians to have continued mining until the white men dispossessed them. Such, strange to say, was not the case. The Chippewas, who occupied the South Shore when the Jesuits and later English explorers came, had a few copper articles but they knew nothing of mining. Trees that were found growing in the ancient workings when these were rediscovered in 1847 proved that at least three hundred and fifty years had elapsed since the last work was done. The height of the ancient Lake Superior copper industry is estimated to have been reached over five hundred years ago. Soon after, long before the Europeans penetrated as far west as Michigan, it stopped completely. The end came suddenly and unexpectedly. Tools were left in the pits all ready for continuing work, and masses of copper were half pried out of the rock. What calamity ended this comparatively civilized period we do not know. Perhaps the hardier more savage Chippewas annihilated the workmen by a surprise attack. Or maybe a pestilence or a famine drove them out, never to return. Whatever the reason may have been, the first Copper Country miners completely disappeared. A dense forest grew over the ancient workings. So thoroughly were the traces of the earlier mining hidden by trees and brush that the district had to be discovered all over again when the white men finally started their mining. The ancient workings were not found until four years after the first modern mines were opened.

Like a lot of great enterprises the development of the Copper Country by white men started with a harebrained adventure and a stock swindle. Cartier and Allouez and the other Jesuit explorers were a credulous lot, and they included in their reports all the tales of mountains of silver and copper that they picked up from the Indians. The stories filtered into England, growing as they traveled. When the French and Indian wars finally won the great north country for Great Britain, many English capitalists were eager to back the plans of adventurous promoters who thought they could cash in on the treasures the French had been too slow to gather.

Captain Jonathan Carver loved adventure, and he could tell a good story. He polished up Jacques Cartier's tale of the mountain of silver and copper on the south shore of Lake Superior until no one could resist it. A band of young Englishmen as adventurous as himself agreed to join him in an expedition to the wonderful mountain, and London capitalists financed the treasure hunt in return for a chance to get in on the ground floor of the mining company they planned to form. In 1770 Carver and his party sailed up Lake Erie and Lake Huron to the Soo Falls in the St. Marie River. There the unknown wilderness began. They packed their guns and blankets and a few trinkets to trade with the Indians in birch-bark canoes, and with friendly Indians as guides paddled along the south shore of Lake Superior.

It was not the pleasure trip they had hoped for. The lake was big and cold, and sudden storms often threatened to upset the fragile canoes. A dense pine forest formed a trackless wall behind the rocky shore. Captain Carver and his fellow-adventurers paddled up a few streams that flowed into the lake, but they did not dare strike off into the forest for fear of getting lost or of being attacked by the hostile Chippewas. If the mountains of silver had been there the treasure hunters would have had little chance of finding them under the dense covering of forest and swamp and gravel. Exploring was a tough job, they found. Still they hated to turn back. Slowly they worked their way along the shore for hundreds of miles, clear around the Keweenaw Peninsula that sticks far out into the center of Lake Superior. On the far side of it they paddled for thirty miles up a little winding river overhung with trees. The Indians called it the Ontonagon. There they did find a few pebbles of copper, and the guides told them that now and then they had found copper pebbles with silver on one end.

This was a far cry from the mountains of silver and copper Cartier had told about. However, it was better than nothing. The maples were turning yellow with the first frost, and they did not dare go farther for fear they might be caught by the terrible winter gales that would be sure death in their birch-bark canoes. They had enough of pioneering, anyhow. So back they went to London as fast as their paddles and sails would carry them.

As they got farther from the wilderness their imaginations worked a magic change in the little copper showings they had found. The pebbles they brought back became only samples of great copper ore bodies waiting to be mined. The copper would surely give place to silver as they developed it. By the time they reached home the story was so convincing that it would pry money loose from the most wary investor. They organized a company to explore the new deposits and sold stocks to the hundreds of stay-at-homes who wanted to get rich quick without suffering the hardships of the pioneers.

Early in the next spring the new company sent out a party of Cornish miners to open up the promised ore bodies. Captain Carver himself did not want any more pioneering. He sent a miner named Alexander Henry as leader of the new expedition. After months of traveling by sailing ship and canoe they pitched their camp on the bank of the Ontonagon. The Indian guides showed them where Carver had found the copper boulders the year before. It looked good to Henry. In Cornwall, copper in the surface gravel meant a good vein close by. The new district was a long way from home, but the same rules would surely apply. A few feet of development was all they had to do. So they made a little clearing in the pine wilderness, built rough log cabins like those they had seen in their slow journey from the sea coast, and started a tunnel into the ridge that should contain the ore bodies.

All through the bitterly cold Michigan winter they drove the tunnel. And every day it looked less promising. Foot after foot the tunnel stayed in clay and gravel. Occasionally they found another copper pebble in the clay, but this was nothing they could mine. It began to look as though this strange country didn't have any bedrock at all. There was nothing like this in Cornwall. If they could have really got underground with solid rock to drill in, it wouldn't have been so bad, for any hard rock mine is home to a Cornishman. But this everlasting frozen clay in the tunnel and the bitter cold wind outside got on their nerves. There wasn't enough to eat either. They had traded their beads and cloth for Indian corn, but the store was getting low. And Cornish miners were a lot better at hitting a drill with a hammer than a deer with a bullet.

On top of all the trouble, with the first warm days in spring, the frozen gravel thawed and their tunnel caved tight. This was too much. They would be "bloody well blowed" if they were going to start all over again timbering their bloody tunnel when there probably wasn't any decent bedrock in the country anyhow. As far as they were concerned, she was "deep enough." So they packed up their scanty belongings and started for home. As it turned out, it was lucky they did not keep on. For although their tunnel was near the Victoria Mine, that later produced a large tonnage of very rich copper ore, the amateur geologists who started the work had done a poor job in locating it. Henry would have had to drive it about a mile through clay and sandstone before it entered the copper formation.

Captain Carver's venture hardly made a dent in the Lake Superior wilderness. All it did was to discredit the earlier reports of the Jesuits. Still the legend that the Indians had worked copper mines in northern Michigan managed to persist. As the years passed, the legend grew into a rumor that seemed reliable. Legends have a habit of doing this. Finally it became so strong that in 1800 Congress passed a resolution to investigate these tales of ancient mines. This buried the question for another forty years.

Even Congress could not kill the tenacious rumors completely. A generation of romantic spirits insisted on believing in spite of the negative reports. Michigan became one of the frontier states. The first governor, Steven T. Mason, was anxious to have his full complement of officials to add to the dignity of his office and to the number of his political followers. Among the positions he could fill was that of State Geologist. One sort of scientist must have looked about like another to Governor Mason. A young botanist and physician named Douglass Houghton had been interested in the development of the wilderness that then made up the northern peninsula, and had been a member of an exploring party along the south shore of Lake Superior. Incidentally he was a "deserving Democrat." This was qualification enough, so Houghton was appointed State Geologist.

In spite of the misfit title, the appointment was probably the best one ever made in Michigan. Douglass Houghton was a keen and enthusiastic observer if he was not a trained geologist. In 1840 he made a reconnaissance of the Keweenaw Peninsula. And strange to say, he found that the rumors of rich copper deposits had some basis in fact. He did not actually find the old workings, but he did find outcrops of native copper. His 1841 report stated that he felt sure the northern peninsula contained copper in quantities that should be commercial. This report was the real beginning of the copper industry of the Lake Superior country. Houghton followed up his first visit by several others. New discoveries made him more and more certain that there was a great copper district in the Keweenaw Peninsula. He was hunting again for the big ore bodies in 1845, when he was drowned in Eagle River. The truth of his optimistic prophecies had not yet been demonstrated. But he was responsible for the development of the Copper Country. The chief city in the northern peninsula was named for him, and Douglass Houghton has been looked up to ever since as the father of copper mining in the United States.

As soon as Houghton's first report was published, prospectors began to filter in to northern Michigan. It was no easy task even to get there. The shores of Lake Superior were still a wilderness, unbroken save for the occasional game trails of the Indians and the trappers. Those who tried to come to the new copper country overland, from the prairies of southern Wisconsin, had to pack their supplies on their backs for weeks of hard tramping. Many never made it at all. The water route to the Keweenaw Peninsula was nearly as difficult. A fur-trading company had built a little canal around the falls of the St. Marie River forty-five years before. In the War of 1812 the locks had been destroyed and were never rebuilt. So the prospectors of 1843 had to travel much as Alexander Henry had done seventy years before, following the south shore of

Lake Superior in rough skiffs or birch-bark canoes. The prospectors of that day had to be woodsmen and boatmen as well as miners.

Jim Paull and Nick Miniclear were the first of the twenty copper hunters who came to the South Shore in 1843, just after the Chippewas had ceded thirty thousand acres of possible mining land to the government. They paddled along the shore of the lake for weeks. Wherever rocks were exposed, they stopped to prospect. At length they came to the Ontonagon River, where Jonathan Carver had found the copper pebbles seventy years before. Paull and Miniclear followed the river a little farther from its mouth. There, in a little clearing, half covered by blueberry bushes, they found a big boulder of pure copper that weighed five or six tons. It was lying on top of the ground, far from any rock outcrops. On its corners were scratches and scars where the prehistoric miners had chiseled off chips of copper. Paull and Miniclear did not stop to find out that the nearest possible source was two miles farther up the river. Copper was worth twenty cents a pound, and their boulder meant a nice little fortune for two "Cousin Jack" miners. So they built a raft and floated the copper down the river on the way back to civilization.

They did not get far. At Copper Harbor, near the end of the Keweenaw Peninsula, the government had just established a mining and land office. The agent discovered that Paull and Miniclear had found their copper boulder outside of the area ceded from the Indians. In order to show his authority he confiscated the copper and would have thrown the two prospectors in jail if there had been any jail. As there wasn't room to keep them in the two or three log cabins that made up the village, he had to be content with kicking them out of camp.

One place was about as good as another to prospect in. Paull and Miniclear wandered back along the shore of the peninsula, hunting for deer and copper as they went. At least they knew there was real copper in the country. Near the northern point they stumbled on the first workable copper that white men had found in place in the rock. It was only a little seam of copper oxide. The two miners scraped together a few tons of it and shipped it by boat to the eastern states.

This is all history has to say about the first white copper miners in the Lake Superior region. They not only failed to profit by their discovery, but they nearly landed in jail to boot. They were probably glad to go back to the Pennsylvania coal mines, where they could at least make a living without having to fight the wilderness.

Luckily some of the other prospectors were more successful. A group of them formed the Pittsburgh and Boston Mining Company in 1844 to sink a shaft on a vein they found near Copper Harbor. They spent $25,000 and took out $2,968, which was not so good. The next year the same company started another mine they called the Cliff, because the vein outcropped in a bluff near Eagle River. The new shaft ran into large masses of copper, and in 1849 the Pittsburgh and Boston Company paid the first dividend from the Copper Country. A full-blown mining boom soon followed. By 1850, twenty-five or more development companies were in the field. Their hopes were high but in most cases bank accounts were low. Few of the companies were successful. But now and then one of them found a rich vein. The development they started proved that northern Michigan was one of the world's greatest copper districts.

Fortunately the really great mines were the last ones that were found. The copper business of the United States was very much of an infant in those days. In the seven years after the Pittsburgh and Boston started shipments, the whole country produced only 5,000,000 pounds of copper, of which 80 per cent came from Lake Superior. A great modern mine produces 5,000,000 pounds in a week – when it can sell it. Even one moderate sized deposit in the early years would have flooded the market with so much metal that no copper miners could have made a living for ten or fifteen years.

The strange quirk of fate that so often supplies natural resources just when they are needed, prevented this calamity.

The Cliff Mine, shown in these pictures, was one of the oldest in the Lake Superior Copper Country. Shut down many decades ago due to exhaustion of its rich vein of Mass copper deposits, it is now owned by the Calumet and Hecla Division of Universal Oil Products Company, which is under lease to Homestake Mining Company and International Nickel.

It arranged to have the poorest copper deposits developed first and the best ones last. The production just kept pace with the demand.

The little vein of copper oxide that Jim Paull and Nick Miniclear worked in 1843 was the only one of this sort in the district. The first ore body in the Copper Country was hardly more than a curiosity.

The next discoveries were "mass" deposits in cross veins. These are fissures or cracks cutting across the rock formation. They are usually only a few hundred feet long, and very irregular. For long distances they are tight seams with only occasional pellets or films of copper. Then they suddenly open up to a width of five or ten feet of rich ore. The copper occurs as the pure metal, sometimes in small particles but often in large masses from which ore bodies of this type take their name. Some of the copper masses are ten or fifteen feet long and several feet thick, containing several hundred tons of pure copper in one solid chunk. Such masses were an embarrassment of riches before the invention of compressed-air drills and acetylene torches, as the copper had to be laboriously chiseled off by hand before it could be sent to the smelters. Pieces weighing only a few pounds were of course ideal, for then pure copper could be sorted out of the veins and melted down into ingots at insignificant cost.

All of the veins discovered up to 1850 were of this Mass type. The first of them were the Cliff and other mines near the north end of the Keweenaw Peninsula. Then in 1847 S. O. Knapp noticed curious depressions in the forest seventy-five miles farther southwest, near the Ontonagon River. He started to dig and found the old Indian workings, with rich copper masses waiting to be mined. This find resulted in rapid development. Knapp's Minnesota Mine was soon paying large dividends. The ore in the successful Mass mines was so rich that it yielded fine profits. But the ore bodies proved to be scattered through much greater areas of barren vein. Out of twenty-four mining companies formed between 1844 and 1850 only six paid any dividends at all. And in the seventy-six years from 1849 to 1925 the total dividends paid by all the mass mines amounted to less than $7,000,000. Altogether operations in the cross veins have resulted in a great loss. The rich bunches of easily found copper in these ore bodies were responsible for the development of the Copper Country. And they supplied all the copper the country could use in the decade before the advent of the Industrial Age. But if the mass mines had been the only ones in the Lake Superior district, the Keweenaw Peninsula would have remained a wilderness broken only by short-lived camps of a few hundred miners.

With the Civil War and the increased use of machinery that followed it, the uses of copper grew so rapidly that consumption doubled every four or five years. The price soared to forty-seven cents a pound in 1864, and stayed above twenty cents for about thirty years. These figures would have come like a breath of heaven to the hard-pressed copper miners of the winter of 1932, struggling to keep alive with five-cent copper. If the Mass mines had been the whole story in the Copper Country, copper would have become almost a precious metal in the 'sixties and 'seventies. The era of electricity might have been postponed for many years. Fortunately the Lake Superior miners stumbled on far greater ore bodies of an entirely different type. They came just in time to bring great fortunes to their stockholders and to give the nation the metal that was essential to its progress.

The new discoveries came slowly. In the 1840s, Louis Agassiz and other visiting scientists recognized that the Keweenaw Peninsula was made up of old lava beds with a total thickness of many thousand feet, separated by thinner beds of sandstone and of "pudding-stone" or conglomerate that had been laid down during intervals between eruptions. In the millions of years since the volcanoes became extinct, the surface of the earth was crumpled and twisted until the beds of lava, once horizontal, were tilted to an inclination of twenty to fifty degrees to the northwest. The veins that contained the Mass copper deposits cut across these tilted lava and conglomerate beds. Instead of being uniformly distributed, most of the copper in these veins occurred where they cut a certain sort of lava, in which bubbles of gas had been imprisoned under a hardening crust. These gas cavities were later filled with various minerals. From the oval shape of the bubbles, such lava beds were called "amygdaloidal," from the Greek word for almond.

As early as 1851 prospectors noticed that now and then there were films of copper around the edges of the white or pink crystals that filled some of the amygdaloidal cavities. The Quincy Mining Company tried to mine one of these beds in the central part of the district, near Portage Lake and the town of Houghton. The attempt was unsuccessful, as this ore was much too lean to work at a profit. But a few years later, in 1856, another exploration company called the Pewabic found near by another lava bed that was much richer. The Pewabic and Quincy companies combined, and started to mine this new ore. It proved to be leaner than that in the Mass veins, but it was much more regular. And the ore bodies were many times as great. They grew with development until they were many thousand feet long and often twenty feet wide. Shafts a mile or two deep showed no decrease in grade.

The Amygdaloid lodes, named from the lava beds containing almond-shaped cavities in which they occurred, had at last made the Copper Country a really great mining district. Hoisting machines on a scale never before dreamed of were designed to lift thousands of tons of ore a day from thousands of feet below the surface. Great stamp mills were built to crush the ore so that the seams and pellets of copper could be separated from the barren rock before they were smelted. The wilderness gave way to attractive little cities that housed the thousands of miners and mill men. Settlers cleared away the forest and started farms to supply the new communities with food. In 1857 a ship canal around the St. Marie Falls solved the problem of cheap freight to the cities of the East. Railroads came only a little later. In a few years the Keweenaw Peninsula became one of the great industrial centers of the country.

Other lava beds proved to be just as rich as the Pewabic. The greatest of them all — the Kearsarge Amygdaloid lode — has been mined for an almost continuous length of seven miles, from one to two miles down the dip. It has yielded over $50,000,000 in dividends. Found in 1879, this is still one of the most important veins in the Copper Country. The famous Baltic Lode of the Copper Range Company was found in 1882 but it did not become an important producer until twenty-five years later. It was so rich that more than $40,000,000 in dividends were paid from it in about fifteen years. Next to Rio Tinto, the Amygdaloid mines of the Lake Superior district were the first really great copper mines found in the world. Altogether they have paid $150,000,000 in dividends.

Even the Amygdaloid ore bodies were soon too small to keep up with the country's appetite for copper. Electricity was coming into use. And every piece of electrical equipment must be largely made of copper. To bring the inventions of Edison to the growing millions in the United States, the Copper Country had to make one more effort. It came through with still another type of ore body. The new "Conglomerate" lodes were as much greater than the Amygdaloids as the Amygdaloids were greater than the Mass veins. They came just at the right time to put the price of copper down to fifteen cents, so that the manufacturers of the new electrical apparatus could make their equipment at a reasonable cost.

No one knows exactly how these wonderful new copper deposits were found. Two stories are told, and the advocates of each story are equally certain theirs is the true one. A hard-working surveyor named E. J. Hulbert is the hero of both accounts. The first tale is that one day in 1861 when Hulbert was running a road survey he stopped to rest at the rough cabin of a squatter who had made a little clearing and was trying to start a farm. The squatter was greatly exercised over the disappearance of his pig — the most valuable possession he had in the world. After a time he heard faint squeals from the dense forest that surrounded the clearing. Hulbert and the squatter followed the squeals until they came to a thick clump of blueberry bushes. The squeals came from the center of the clump. Working their way through the bushes, the two men almost fell into a ten-foot hole, entirely hidden by the leaves. And in the bottom of the hole was the lost pig, who had fallen in and could not get out.

Hulbert helped the squatter to rescue the involuntary explorer, and then climbed down himself to see what sort of a hole it was. And in the bottom he found hundreds of pounds of copper, in chunks of all sizes. Tool marks showed that he had discovered an old Indian cache, forgotten for hundreds of years. And stranger still, the rock in the bottom of the hole was high-grade ore unlike anything Hulbert had seen. It was a pudding-stone, or conglomerate, with the pebbles held together by pure copper. The pig had led Hulbert to the greatest copper ore body that had ever been found.

The other story gives Hulbert all the credit. According to it, he happened to notice a boulder of conglomerate cemented by copper while he was chopping out a line for the road survey. He recognized that the conglomerate was from one of the beds that lay between the flows of lava, and that it was a new sort of ore body. For three years he spent all his spare time hunting for the new lode. He found the Indian cache, but according to this story the old pit was in barren rock. It was only on September 17, 1864, that a prospect hole he sunk through the glacial drift broke into the conglomerate lode, and the Calumet and Hecla ore body was at last discovered.

The distinguished engineers and mine owners who knew Hulbert in the years of success that followed accept the second version of the discovery. Their desire to honor their friend and to give full credit to engineering skill may have something to do with their choice. Any true lover of romance must lean in the opposite direction. As Rafael Pumpelly charmingly says in his autobiography, "I like to believe that credit was fundamentally due to the pig."

Whether the pig or science was at the bottom of it, Hulbert had to have money before he could develop the ore bodies. Some lucky star led him to just the right men. Why he went to Boston is not known. For Boston was then a sober and conservative city, as far from a mining center as any place that could be imagined. And it was to the very heart of old Boston that he penetrated. Some spark of red romance must have lingered through the long generations of blue blood in the Pilgrim families. Hulbert's story appealed so strongly to this buried spirit of adventure that the first directorate of the exploration company he formed in 1861 read like a passenger list of the *Mayflower*. Quincy Adams Shaw and Horatio Bigelow headed the list, with Higginson, Hunnewell, Livermore and other famous New England names following them. Their money and faith in Hulbert's ability made Calumet and Hecla.

The development of the Conglomerate ore was delayed a few years by the uncertainties due to the Civil War. The Hulbert Mining Company had bought 1,920 acres of government land and could afford to wait. Hulbert himself kept up his surveying business, but spent his spare time studying the new lode. At last he started the first new shaft in the conglomerate ore body in the fall of 1864.

This was the famous Calumet Number I Shaft. It followed the ore down the dip of the conglomerate bed. The lode was twenty to thirty feet thick, and wonderfully rich. Every foot of development added hundreds of tons to the amount of ore that could be measured. Hulbert's knowledge of the geology helped him to follow the course of the bed along the surface, under the soil and gravel. Other shafts quickly entered ore like that first discovered. One of them, later called the Hecla Number I, was only a few feet from the original Indian cache. Whether or not the Indians had found the ore, their storage hole was on the edge of one of the greatest copper ore bodies in the world.

In 1865 Quincy Adams Shaw advanced $16,800 to buy more land for the Calumet Company. This completed the area that was later found to contain the entire outcrop of the ore body. Hulbert had studied the ground so carefully that the only chance he left for others was to sink half a mile to cut the downward extension of the ore.

A few months' development proved beyond question that the Calumet and Hecla Conglomerate was a magnificent ore body. But dividends were still several years in the future. Stamp mills, concentrating plants and a smelter had to be built before production started. These required money. A five-dollar assessment was made on the 20,000 shares of Calumet stock. This was all right for the wealthy Boston stockholders, but it was a hard blow for Hulbert and his hard-working friends in the Copper Country. Other assessments followed in rapid succession until $12.50 had been called. In all $1,200,000 was spent before the two companies were ready to produce copper. Hulbert could not pay his assessments, as he owned over 50 per cent of the stock of the Calumet Company and nearly as much of the Hecla. He was therefore forced to sell nearly all his stock at a price that he knew was far less than the real value. This nearly brought on a legal fight between him and the Boston officials. Fortunately they appreciated his services and finally gave him $300,000 in paid-up stock. Dividends amounted to many times this amount. Hulbert was one of the few discoverers of great mines who received his fair share of the profits.

Another famous New England name was added to the Calumet and Hecla list in the development and equipment period. The great naturalist Louis Agassiz had visited the district in 1848. His visit may have had something to do with interesting the Boston capitalists in copper mining. Nineteen years later his son Alexander Agassiz became superintendent of the new Calumet and Hecla mines. He acquired enough stock to make him in a few years one of the wealthy men of the country. Alexander Agassiz was actually in charge of operations in the Copper Country for only a few years. But his management was so successful that a little later his Boston associates made him president of Calumet and Hecla. Under his direction it became the copper company with a Harvard

Edwin J. Hulbert, above left, found the great Calumet and Hecla Conglomerate Lode, with the help of a pig, in 1861. Alexander Agassiz, right, made C. and H. for many years the world's greatest producer of copper.

Below, the Red Jacket Shaft of Tamarack Mining Company, shown with its crew about 1885, was the first Northern Michigan copper mine to exceed 5000 feet in inclined depth. It was later part of the Calumet and Hecla.

accent – a source of wonder and envy to all the rough and hardy western copper miners for many decades.

Hecla Mining Company paid the first dividend from the great Conglomerate ore body in December, 1869. Calumet Mining Company followed with a dividend six months later. By the middle of 1871, when the two companies were consolidated to form the Calumet and Hecla Mining Company, $2,800,000 had already been paid. From that time on millions were piled up at an incredible rate. On the $1,500,000 investment, total dividends of $152,250,000 were paid between 1871 and 1921 – a return of over a hundred for one. Twenty-eight additional millions have been paid since the Calumet and Hecla absorbed several other properties in 1923. The stock on which $12.50 was collected in assessments sold during the war for $1,000 per share. The rapidly mounting dividends made Boston the center of the world for copper finance.

In making its crop of new millionaires, Calumet and Hecla did not forget the miners who made the success possible. It built attractive towns and rented the houses to the employees at low rates. The streets of Calumet, lined with trees and flowers, are in striking contrast with the bare and desolate camps of the mountain states. Wages were always high, and working conditions as good as the very deep mines allowed. "C. and H." created thousands of contented homes in the Copper Country in addition to the palaces of the Boston stockholders.

In the years that followed the discovery of the Conglomerate Lode dozens of new companies attempted to duplicate the success of Calumet and Hecla. Eight of them paid dividends of many million dollars. They would have been thought phenomenally successful if they had not come so late that they were dwarfed by Calumet and Hecla and by the giant companies of Butte and of Arizona. Many more of the explorations fell by the wayside, or found just enough ore to lead their stockholders on to throw new dollars after old. Yet the proportion of successful mines in the Copper Country has been greater than in any of the other copper districts. Out of a hundred companies that have started development, twenty-eight have paid dividends. This is a remarkable record in a business in which one success in ten is all that can usually be hoped for.

Hulbert and his successors in Calumet and Hecla made only one mistake when they bought the ground to protect the big Conglomerate ore body. They acquired the full length that has since been productive – about two and a half miles along the lode. For most of the length they owned so great a width that no matter how deep the ore continued on the 25-degree dip, it would stay within their property. But northwest of the center of the outcrop there was a piece of ground half a mile long that they did not buy. It was so far from the place where the ore came to the surface that its value seemed very problematical for the first few years. The big lode would not pass into this foreign property until it extended thirty-five hundred feet down the incline. Only a handful of ore bodies in the world had been worked at a profit at that depth, and at the time it seemed too optimistic to hope that the Calumet and Hecla Conglomerate would add another to the list. Even if the ore did continue to the adjoining property another company would have to sink a vertical shaft nearly half a mile through hard, barren rock to develop it. This would cost millions. The Calumet and Hecla directors thought they were safe in postponing the purchase, at least for a few years.

As the big mine grew from year to year, the prospect that ore would continue to great depth became much brighter. The stopes were less than a mile long on the upper levels, with a 3,000 foot barren area in the center. On every level the ore became longer and the barren area smaller. In 1880 the twenty-fifth level was developed. It cut almost continuous ore for 8,500 feet. The property to the northwest looked attractive then. But the price had gone up so far that it was still a doubtful gamble. Prosperity had made the directors conservative, and they decided not to take so expensive a chance.

Before they could change their minds, it was too late. One of the old Copper Country mine foremen, Captain John

Daniell, saw his great opportunity to make a fortune. He was sure the C. and H. Conglomerate would extend another thousand feet down the dip and that the old company was overlooking many millions in profits. So he followed the trail of Edwin Hulbert back to the Boston bank accounts. The Bigelows and J. W. Clark of the original Calumet and Hecla group were persuaded by his enthusiasm. They formed the Tamarack Mining Company in 1881. With $1,500,000 capital, they bought the 1,120 acres of open ground and started to sink a big vertical shaft half a mile deep to cut the projected course of the lode. For more than three years the sinking went steadily ahead. Every foot of the shaft took sixty-one dollars out of the treasury. It was poker with the highest ante on record, but the pot was worth the risk. Captain Daniell, like all Cornish miners, looked with scorn on "them damned geologists." But fortunately for the Tamarack stockholders, he could figure dips with the best of them. He told his backers that the conglomerate lode would be cut at 2,260 feet vertical depth. On June 20, 1885, the shaft entered rich ore at 2,270 feet – just ten feet more than Captain Daniell had estimated before ground was broken.

Success was then assured. It was still necessary to drift along the vein, open up deeper levels for mining, and build a mill. But this was easy when every round blasted in the drifts opened up more high-grade ore. The first dividend was paid in 1888. In twenty years the stockholders received more than six times the original capital. Then the mine was becoming so deep that expenses rapidly increased. Five shafts were required to mine the ore, the last of them over a mile deep vertically. At this great depth ventilation was bad, and the efficiency of the miners low. The drop in the price of copper in 1907 ended the dividends. For several years the company struggled along, sometimes making a little and sometimes losing. The Calumet and Hecla, with its many shafts and fine surface plant, could mine the ore much more cheaply than Tamarack could. Finally in 1917 the Tamarack property was sold for $3,600,000 in Calumet and Hecla stock. The total

Captain Jack Daniell, right, got the deep part of the Conglomerate Lode with Tamarack's Red Jacket Shaft. The picture of underground workings in that mine, sometime prior to 1900, shows the large timbers needed to support the heavy ground. (*–Right,* THE DAILY MINING GAZETTE, *courtesy M.T.U. Library)*

Workings in the Tamarack Mine demonstrate methods of hauling ore and drilling by "double Jack" hand hammers before the time of compressed-air drills.

return on the $1,500,000 investment was a little more than $12,000,000. Captain Daniell had won the big pot.

The end of the Gay Nineties brought the last of the great discoveries in the story of the Copper Country. The rich ore bodies in the Mohawk Mine and in the Baltic Lode, later owned by the Copper Range Company, were found at about the same time – in 1898 and 1899. Calumet and Hecla reached its peak of dividends by paying $10,000,000 in the latter year. It looked as though the old mines were getting better, and the prospect for many new ones looked bright. All the mining prophets foretold many decades of increasing production.

But even the best mines must come to an end, and the Copper Country had already started downhill. That is the sad part of mining. Every ton of ore taken from the ground leaves just that much less for the future. The decline really begins the day production starts. It can be concealed for a time, first by actual new discoveries, and then by including in the reserves low-grade material that is not rich enough to yield a profit in the early days of the mine. The hopes aroused by these apparent gains blind the stockholders for a time to the fact that their assets are waning. At last they wake up to find that their ore reserves are no longer ore. The best they can do is to drag on for a few years, teased by small profits in the good years, until a time of low metal prices brings the inevitable end.

The downhill path of the Copper Country has been a long one. Happy intervals of prosperity have broken it. The boom and the high copper price of 1907 brought forth a flock of new companies with bright prospects. War prices again started all the old mines working full force. The peak of production in the whole history of the district was reached in 1917, when Calumet and Hecla alone boasted $8,500,000 in dividends. This prosperity was short lived. The mines were fast approaching two miles in inclined depth, and costs were mounting. The end seemed near in the lean years from 1921 to 1924.

The post-war boom again brought a few prosperous years. Calumet and Hecla made 90,000,000 pounds of copper in 1929, and paid $9,000,000 in dividends. This was only a flash of the old glory – maybe the last one. The depression years that followed filled the Copper Country with idle miners. The company balance sheets made sad reading in those days.

The old district would not die without a struggle. The mines still contain 100,000,000 tons of one per cent ore. Unfortunately this ore occurs a mile below the surface. Even the best management could not make low-cost copper out of lean ore at that depth. While Michigan costs were mounting, new copper districts in Canada, Chile, and South Africa developed many hundred million tons of 3 and 4 per cent ore in enormous bodies close to the surface. It seemed hopeless for the old Copper Country to try to compete with these new mines.

The officers of the Michigan companies quickly realized this. They had one chance of avoiding the fast approaching

This village was at the abandoned Cliff Mine, now owned by Calumet and Hecla.

ruin. If the cheap copper from Canada and Chile could be kept out of the United States, the domestic market might keep them alive. In 1930 they took the lead in a campaign for a tariff on copper. All authorities said the fight was hopeless. But in May, 1932, Congress enacted a four-cent import tax and the President signed it.

Cheering crowds and torch-light processions greeted the news. In the Copper Country towns prosperity might be a long way off. But at least they could hope. The Calumet miners would have to buckle their belts a notch or two tighter, but some day the great hoists would again be lifting ten thousand tons of ore a day from a mile below sea level. The profits may not be great, as it will be a hard struggle to compete even with other American mines. As long as hope is left, this will not matter. Every good Copper Country miner knows that another C. and H. Conglomerate lies hidden beneath the mantle of glacial drift. Any day a diamond drill may break into a bonanza that will make Michigan once more the richest copper district in the world.

AUTHOR'S NOTE: The prophecy at the end of the *Romantic Copper* chapter on the Copper Country was overoptimistic as far as the timing of a comeback was concerned. The outline of what actually happened in the forty years since the great Depression, near the beginning of the second part of this book, makes this clear. In the long run, however, the hopes may still be realized.

Headframe and houses were part of the Mohawk Mine, later acquired by C. and H.

In the photograph opposite above, a loaded ore car waits on the cage in the Tamarack Mine while a miner signals the hoistman. At another C. and H. mine, this hoist with a great conical drum brought up ore from more than a mile inclined depth. The underground photograph below, taken at least fifty years ago, shows the method of timbering at Calumet and Hecla.

A good view of Butte Hill could be had from the Hennessy Building: above, looking east; below, toward the northwest. These pictures show many of the great Butte mines before the large open pits were started.

4

The Richest Hill on Earth

IN the excitement of the last year of the Civil War, only a handful of people paid any attention to two of the most important events in the history of copper mining in the United States. The first, as we have seen, was the discovery of the Calumet and Hecla ore body. Even more unheralded was the birth of the greatest copper camp of them all—Butte, Montana.

It seemed a poor year for mining in general. The crest of the gold excitement in California had gone by. Thousands of prospectors who had failed to make their pile in the trail of the Forty-Niners scattered all over the western mountains in search of new bonanzas. Only a few of them had any luck at all, and none of the discoveries seemed to amount to much. In 1863 the discovery of placer gold in Bannock turned the stream of gold hunters to western Montana. The next year a party of emigrants found gold at the foot of a long yellowish granite hogback at the north edge of the Summit Valley, just west of the Continental Divide. They called the new camp Butte, from a gray volcanic cone that is the most striking landmark in that part of the Rockies. Prospectors flocked in and started to wash gold from the dry sand washes. Their hopes were big and their talk was bigger. But they little thought that the wind-swept ridge that looked down on their discovery was destined to become the Richest Hill on Earth.

A miniature gold boom made Summit Valley for a few years one of the riotous, happy-go-lucky camps that still give Montana an aroma of romance and disrepute. The gamblers and red-light district flourished, and a few placer miners made a stake that carried them on to the next hoped-for bonanza. A million and a half dollars' worth of gold was washed out of the gravels by '66. Then placer mining quietly died. It had not been much of a success, as a gold rush. But it served to call the attention of the prospectors to the quartz veins in the Big Hill.

The California miners had learned that the placers were only half the story. Upstream from them were the quartz veins from which the gold had been washed by ages of frost and wind and rain. So while the first comers in Butte were washing out the placer gold on the flat, those who were a little too late to locate placer ground worked up the gulches with their gold pans to find the source of the gold. For once the search was an easy one. Crossing the long ridges north of camp, they found a network of big veins. The gold content of the veins was low. But some of them contained rich silver ore right at the surface. And one of the veins had a little copper carbonate with the silver. Silver and copper were hardly worth looking for at that time on the far edge of the frontier. Still the miners hoped that gold values might get better below the surface. They located mining claims on the veins that looked most promising and called them according to their fancy with names that became famous in mining history — Parrot, Original, Gagnon, Neversweat, Anaconda and hundreds of others.

The prospect holes sunk on these first lode claims in Butte were disappointing. The gold did not increase as the miners hoped. What to do with this ore was a doubtful problem. The first transcontinental railway was just being built. Before it reached Utah the miners had to haul their ore and supplies by wagon many hundreds of miles across the Indian- and bandit-infested plains between Butte and Fort Benton on the Missouri River. When the Union Pacific had been completed to Ogden, the wagon haul was a little shorter. Still Butte was four hundred miles from the railroad, and freight rates were extremely high. By the time a prospector packed his ore down from the hill, freighted it to Corinne and then paid the railroad to haul it to Colorado or New Jersey smelters it was nearly worn out. There was no hope of profit unless the ore was phenomenally rich.

In spite of the expense of working in so remote a camp, Butte enjoyed a little silver boom in the late 1860s. Mike Farlin, who had staked the first quartz claims in Butte, ran into high-grade silver ore in his Travona Mine in '66. He tried to treat the ore in a small arrastre, by the process that had been developed by the Spaniards in Mexico and later adopted at the great silver mines of the Comstock Lode, in Nevada. The process would not work on Butte silver ores. Little copper and silver smelting furnaces did not bring any better luck. Farlin and the other miners struggled along for a year or two breaking their hearts by paying all the values in hundred-dollar ore to the teamsters and railroads and smelters. It looked hopeless. By the end of 1868, Butte was about ready to join the dismal ranks of the abandoned mining camps. The only miners left were those who did not have money enough to move away and a few enthusiasts like Farlin who did not have sense enough to go.

So far Butte was just like dozens of western camps that, like the desert flowers, blossom for an hour or two and wither away into forgotten dust. But Fate was preparing one of the great surprises that make mining so fascinating a game. The first few years were only the stage setting for the glorious and disreputable drama that was to follow. Now the time had come for the entrance of the first of the trio of actors whose vision was to make Butte great and whose warfare and corruption were to make it infamous.

The entrance of the first of the three Butte stars was not startling. W. A. Clark was just an insignificant looking wandering trader who did a little mining on the side. Starting from a Scotch Irish family in a small Pennsylvania town, he had tried half a dozen occupations in the thirty-three years before he came to Butte. He studied law for a while at Iowa Wesleyan college, taught school for a year or two in Missouri, and did a little placer mining in Colorado and in the early boom

William A. Clark, for a brief period United States Senator, was one of the central figures in the Butte legal fights. *(—Remaining pictures in this chapter are from* A Brief History of Butte, Montana *by Harry C. Freeman)*

in Bannock, Montana. Mining did not bring him much more than wages. The high prices paid for all supplies in the remote camps excited his Scotch blood – if the word excitement can be applied to one so cold and calculating. So he bought provisions and tobacco in Salt Lake, freighted them to the mines, and sold them at a profit of a few hundred per cent. This beat gold mining. By the time he was thirty years old he had accumulated a good stake and was starting a series of banks that would loan money to the miners and storekeepers of the frontier towns for 10 or 12 per cent. He was so diminutive in size and mild and ingratiating in manner that the rough mining communities did not realize what a strong hold he was getting on them. They could not understand a man like Clark at all. In Montana in the 'seventies most men were roughly dressed and rougher in manner, but generous and tolerant of everything save horse stealing. Clark was neatly dressed, delicate in stature and in manner, and prided himself on his culture and refinement. But in business he was treacherous and selfish, and considered any means justified to further his own progress in making money or in politics. He was vindictive, yet had his passions under such control that he would join hands with an enemy rather than lose money fighting him. The simpler men about him were at his mercy in business dealings. Most of them hated him cordially. But he paid men who could be useful to him so liberally that they would fight for him even if they were never sure when he would turn against them.

It was in 1872 that this unsuspected copper king of the future quietly slipped into Butte to see if the silver miners had money enough to make it worth his while to take it away from them. The camp was too dead to offer much prospect for a trader. But the big copper-stained outcrops on the hill attracted his keen imagination. Even if copper ore was not of any value then, it might be someday, when the country was settled. And he could wait if the possible prize was worth waiting for. The prospectors were discouraged, and a few thousand dollars bought Clark several claims along the biggest copper showing. Among them the Original, Colusa, Mountain Chief and Gambetta later became famous.

Clark realized that his training had not equipped him for running big mines. He had that rare and priceless quality – a realization of his limitations. He also knew he could overcome any limitations if it were worth his while. So he went to the Colorado School of Mines to learn all he could about the treatment of copper and silver ores. He was no raw undergraduate, but a mature, ambitious and successful businessman eager to learn. In a year he knew all the professors could teach him and in addition had become so well acquainted with the officials of the Boston and Colorado Smelting Company that they were ready to back him in any project he recommended.

The next year Clark returned to Butte and started to develop his Original and Colusa claims. A few shipments of ore hauled to Corinne and then shipped to the Colorado smelter convinced him that he could not make any money this way. Farlin had also found this out and was trying to build a stamp mill to treat silver ore from the Travona. Clark saw his chance and advanced $30,000 to finish the mill. The mine was the security, and Clark was allowed to work it until the loan was paid back. This was easy for a trader of Clark's experience. The ore suddenly turned low-grade, and as creditor, Clark had to foreclose on the mine to protect his loan. Even Clark's banking partner was discouraged and sold out to him for a bunch of horses. Then the ore became high-grade again. Unkind critics said Clark had deliberately worked in a lean place until he secured the title, but of course this could not be proved. Anyhow Clark now owned both the mine and the mill, and the profits were all his.

At first this mill was the only one in camp, and Clark could charge a rate just under the cost of hauling to the railroad. At $30.00 a ton the profits almost satisfied his grasping nature. But they were too good to last. Other silver mills sprang up, and the cost of treatment rapidly fell. Silver mining became worth-while. In 1877 a shipment to Baltimore of 35 per cent copper ore containing $50.00 per ton in gold and silver had cost so much for hauling and freight that the miners lost money on it. With several mills competing for ore three or four years later twenty-five-dollar ore was rich enough. Before long 290 stamps were noisily grinding 400 tons a day of silver ore. Thousands of miners and bartenders, gamblers and dance-hall girls made Butte the crowning jewel in Montana's diadem of lawless camps.

The cream of profits from silver milling was gone. But Clark had another idea that was a lot better than the first one. In 1879 he persuaded his Colorado smelting friends to build a small furnace in Butte. They called it the Colorado and Montana Smelting Company. Unlike the earlier smelters this one worked. Now Clark could recover copper as well as silver from the rich surface ores in his Original and Colusa mines. The profits came rolling in faster than ever. Soon he was the wealthiest man in Montana. His clothes became more immaculate, his manner more polished, and his ambition more far-reaching. And his conscience was still under perfect control. Already he was looking beyond his frontier state and dreaming of further triumphs at Washington. And a dream was no idle pastime with Clark. It was a fixed goal, to be attained in spite of all obstacles, if need be after many years of work and schemes and plots.

In building up his own fortune, Clark had changed Butte from a dying gold camp to a flourishing silver district. The copper that had at first attracted his attention was only a side issue. It remained for the second of the great actors in the Butte comedy to realize that the silver was hardly more than an insignificant by-product and to build out of copper the greatest mining camp the world had seen.

Nature played one of her best pranks in throwing Marcus Daly and Clark together in one small community. There would hardly have been room for them in one continent. They were born to hate each other. Daly had come to America as a poor Irish boy, with all the Irish qualities of wit and good fellowship and loyalty, combined with an unquenchable love of a good fight. He was rather short and heavy set, with a genial

Marcus Daly was in charge of the properties that became Anaconda after the Butte legal and underground wars.

smile that made friends for him everywhere. And his friends were friends for life, not only for the time they could be of service to him. He was just the opposite of Clark in everything, save the will to succeed. Though not quite as keen mentally, his fine personality gave him a popular appeal that all Clark's money and shrewdness could hardly overcome. Daly's success had been just as rapid as Clark's. He landed in America with nothing in his pockets save his fifteen-year-old Irish smile. A few months' work selling papers and delivering telegraph messages brought him the price of the cheapest ticket by sailing ship to California. The California mines were already past their peak, and Daly's ambition took him on to the new bonanzas of the Comstock Lode in Nevada. A lot of hard work underground, added to his energy and magnetism, soon made him one of the best-known miners in the West. Hearst and Haggin and the other magnates trusted his judgment about a mine more than that of the most highly educated graduate engineers. He was soon making important mine examinations. In 1875 the Walker Brothers, leading bankers of Salt Lake City, sent him to Butte to see if they could make anything out of the new silver excitement. He optioned the Alice Mine for $5,000, receiving a small interest in it for himself. He built a mill and soon made the Alice such a success that he sold his own interest for $30,000.

This $30,000 was just what Daly was looking for. In 1875 a hopeful fellow-countryman of his named Mike Hickey had located the Anaconda and Neversweat claims on a broad band of yellow-stained, crushed rock that formed a bare streak across the slope of Butte Hill. All the experts said he was crazy. That ledge was just broken-up rock and not a vein. Marcus Daly did not care what the experts thought about it. The ledge was a big one, and somewhere in it there must be a lot of ore. So he took the $30,000 and gave it all to Hickey for the claims that so far had only a few shallow prospect pits and a little showing of silver ore.

Daly knew it would take a lot of money to develop his new mine, and it was too big a job for him alone. His old

Comstock employers, George Hearst, Haggin, and Tevis, were the ones to help him. Hearst came to see the prospect before putting up any money. By this time Clark was jealous of the rapid rise of Daly and tried to discredit him with Hearst in every way possible. Daly was only a rough-neck Irish miner, said Clark, and anyone would be crazy to spend money on his recommendation when a lot of the best engineers in Montana said the mine was no good. Luckily, Hearst thought differently, and all Clark succeeded in doing was to acquire a hard-hitting lifelong enemy. In 1881 Daly, Hearst, Haggin and Tevis formed the Anaconda Silver Mining Company and started to sink on the big vein.

At first it did not look too good. There was a little silver ore, but that was all. Daly leased the Dexter mill and treated 8,000 tons of thirty-ounce silver ore that yielded hardly any profit. Then the ore played out. But on the hundred-foot level there was a narrow seam of rich "copper glance" ore. Stuff that ran 30 per cent copper might amount to something, and Hearst and Daly decided to sink again. On the 200 level it was not any better. On the 300 the copper glance was five feet wide. Maybe the Anaconda would be a mine after all. Daly made a few small shipments overland and across the ocean to Swansea, Wales, and started to sink another hundred feet. On the next level the high-grade glance widened out to more than fifty feet. Copper ore of such size and richness had never been heard of. As a silver mining company the Anaconda had been a failure, but it fooled even Daly by becoming the richest copper mine in the world.

There was no question about the success of Butte now. Railroads were built as fast as they could be graded; the Utah Northern to Ogden ran the first trains in December, 1881. Other mines on Butte Hill sunk shafts in the lean iron-stained outcrops and found ore as rich as that in the Anaconda. Four big smelters were soon killing all the trees for miles around with their sulphur fumes. Butte sprawled out over the flat at the base of the big hill so rapidly that it was soon a city of 40,000. And the furnaces turned out an ever-increasing stream of red copper pigs until in 1887, only six years after Marcus Daly found the high-grade glance in the Anaconda, Butte passed the Lake Superior Copper Country in production and became the greatest of all the copper districts.

It was about the ugliest of them all, too. Before the miners came, the wind-swept valley, surrounded by high ridges dark with stunted pine trees, had a desolate sort of beauty. It was not inviting, but the great open spaces and the blue of the distant mountains brought peace to souls tired by too much humanity. There wasn't any peace after the boom began. The stamp mills were roaring day and night. Every few hours the muddy roads leading up to the Big Hill were crowded with swarms of miners black with sulphide dust, going on or off shift. The smelters poured from their tall stacks choking white clouds of sulphur smoke that mingled with the black coal smoke from the chimneys of the houses and power-plants to make a dirty pall through which the sun only occasionally cast dingy yellow shadows. In the winter days there was never any real light — only a dismal dusk that seemed accentuated by the smoke-blackened snow. Everything was drab and ugly. White paint on the one-story shacks the miners lived in became gray in a few hours. And a bitter cold wind blew over the Continental Divide almost constantly, chilling to the bone the sweat-drenched miners as they came off shift. With the cold and the choking smoke and the mud and dirt everywhere it was no wonder that the brightly lighted saloons and dance halls and gambling joints were crowded day and night. The miners had to be noisy and gay to forget the depressing ugliness all around them.

There was nothing depressing about it for Clark and Daly. In addition to the best of the mines they soon controlled banks and stores that got back the pay checks that the saloons and red-light district took from the miners. Millions in profits piled up every year. It seemed as though surely there must be enough for them both, but even Butte was not big enough for them.

Daly had been waiting his chance to get even ever since Clark had told Hearst and Haggin that he was just an ignorant "Mick" miner. His chance came in 1888. Clark's growing ambition led him to try to get elected that year as representative in Congress for Montana. He secured the Democratic nomination without much trouble and without any open scandal. Daly backed the Republican candidate, T. H. Carter, and Clark stood no chance against such a genial and popular campaigner. He was badly beaten in his own city and ward.

This started an eleven-year political war that ended only with Daly's death. The bitterness of the struggle drove men who ordinarily were honest citizens to lengths that seem incredible. In an older, more conservative community they at least would have been less open about bribery and general lawlessness. But in the late 1880s Montana was just emerging from the wildest of the frontier days. The pioneers on whom history throws such a glamour were pretty evenly divided between adventurers and those who had gone West because the sheriff in their home town was too particular. Many of them would cut a throat as casually as they would cut a deck of cards. Until after the Civil War, lynch law and the rule of the vigilantes were the only restraints. As long as a man did not run afoul of them, he could follow his own system of ethics, protecting himself as best he could by his own strength from the encroachment of his neighbors. It was no wonder that in the bitter fights for the mines and for political power the men who had won their way to the front in the pioneer days soon reverted to their old contempt for the due process of law. The only thing that counted was to win. If the law would help them, all right. But if it would not, they were ready to help themselves by bribery or force or any other method that came to hand.

His defeat in the election of 1888 only made Clark all the more determined to succeed in politics. His personality gave him no hope of winning popular votes. But the Montana legislature had only a hundred members, and they elected the United States senators. A few hundred thousand dollars would go a long way toward making the legislators forget a cold and reserved personality.

After five years of plotting, Clark was ready for another big attempt in 1893. The fight was in the open then. All Montana backed one or the other of the leaders. Clark fought with money and influence and Daly with wit and friendliness, backed with a little money where necessary. Each would go to any length to beat the other. When the legislature met, Clark had won all the Democratic members, and six Republicans joined them to bring him within three votes of election. Both sides howled bribery, but if anyone had been bribed it was a good job, for they all stayed put. Finally the legislature adjourned without electing any senator at all; the second round had ended in a draw.

The next year the fight was even more bitter. Daly wanted the state capital moved to the new town of Anaconda, where he had built a big smelter and mill for his Anaconda Company. This would be fatal to Clark, who naturally wanted the legislature to remain at Helena, far from his enemy's center of influence. Each party spent hundreds of thousands of dollars in the capital fight. Clark's *Butte Miner* and Daly's *Anaconda Standard* led the rival armies with burning editorials. This round was won by Clark, and the capital stayed where it was. The decision saved Helena from declining into a fourth-rate little smelter town, and the citizens gave Clark the warmest ovation of his life. They paused in the celebration long enough to bury Daly in effigy. The gain to their pocket-books more than made up for any lack of affability in their hero of the hour.

Four years later there was another campaign for the senatorship. It started with an election of representatives for the state legislature. Party lines were thrown overboard. The only issue was Clark or no Clark. Clark seemed far ahead. But early in the morning following the election two masked men broke into the room where the judges were counting the ballots in Daly's precinct and opened fire with revolvers. In a few seconds they killed one judge, wounded another, scattered the ballots and escaped. This outrage was too much even for the political ethics of Montana in the frontier days. Clark was thought to be at the bottom of it, and everyone assumed he was through as a candidate. Daly went East and abroad for his health, sure he had won the fight.

But Clark was not the sort of fighter to quit while his bank account still held out. The legislature met in January, 1899, amid a flood of rumors of bribery. Fred Whiteside, a lawyer who had formerly supported Clark, induced the two houses to appoint a joint committee to investigate the rumors. The day that followed was like a chapter from a dime novel. A Clark representative offered Whiteside $30,000 for his vote. Whiteside took the money and deposited it in a safe-deposit box, together with a sworn statement that he had accepted the bribe to get evidence for the joint hearing. The bank did not seem safe enough, so he later hid the money and affidavit in his mattress in the hotel. Just before the committee met, Clark's son and four supporters forced Whiteside to come with them to a hotel room. A bitter argument ended when one of them threatened to kill him if he testified. He was wise enough to know that even in a Montana election they would not dare commit quite so open a murder. So he broke away from the Clark agents and ran to join the committee. The committee went with him to his room and received the $30,000 and his sworn affidavit. Other evidence showed bribes of $170,000 more. The legislature could not overlook such open and wholesale corruption and urged the Grand Jury to act on the committee revelations. Next day it gave Clark only seven votes for the senatorship.

Even now Clark would not quit. Just before the legislature started balloting again for the senatorship a few days later, the Grand Jury handed in a decision to the effect that there was not enough evidence to convict Clark's associates of bribery. The legislature promptly unseated Whiteside and elected Clark senator. In two weeks he had persuaded forty-seven of the legislators who had voted against him to change their minds. Among them was Miles Finlen, one of Daly's best friends in Butte. Clark had finally won and all Helena drank his health in $30,000 worth of his champagne.

The victory proved an empty one. As soon as Congress met, the other senator from Montana, T. H. Carter, petitioned the Senate to declare the election void because it had been secured by bribery. Senator Hoar conducted the hearings, and examined most of the leading citizens of Montana. The evidence was almost beyond belief. Clark agents had paid the grand jurors $10,000 apiece for their decision. Bribes to the forty-seven legislators who had so obligingly changed their minds totaled $431,000. Clark had spent as much again in other ways, legitimate and illegitimate. On April 10, 1899, the Senate Committee unanimously and indignantly voted that the election should be declared void.

Still Clark would not give up. The fact that his corruption had been published for all the world to read made no difference. Everything was fair in a Montana fight. The first move was to prevent the Senate from ousting him. It was an old trick for a miner to tell the boss to go to hell five minutes before he was fired. Clark worked the same game and resigned just before the Senate voted on the Investigating Committee report. He made the lame excuse that the bribery charges were all trumped up by Daly.

The resignation left the Montana senatorship vacant, and the governor could appoint a successor for the rest of the term. Unfortunately the governor, R. B. Smith, was on the Daly side of the fence. Clark knew he could not win him over. A shrewd trick and the rumored expenditure of $200,000 got around this difficulty. Just before Clark resigned, Governor Smith left for a short trip to San Francisco. Lieutenant-governor Spriggs was in Sioux Falls, South Dakota, so Montana was left without a chief executive. But Spriggs unexpectedly appeared again in Helena just in time to receive Clark's resignation. With the governor out of the state, Spriggs declared himself acting governor and appointed Clark senator to succeed himself. Clark was out and in again in two days.

Such persistence should certainly have been rewarded. But still the way of the would-be statesman was hard. Governor Smith returned to Montana as fast as the train would carry him and revoked the appointment on the ground it had been made by fraud. The case was never settled, as Clark was content with the title and did not try to take the contested seat in the Senate.

Two years later Montana elected a senator again. This time Marcus Daly was taken sick in New York, on his way from Europe to Butte to re-enter the fight. With the chief enemy out of the way, the 1901 election was easy. Clark was triumphantly "vindicated" and was at last a real senator. He could now honorably retire on his laurels. He never sought re-election, but for the rest of his long life his ornate palace on Fifth Avenue was known as that of the great statesman and Copper King – Senator Clark.

Meanwhile the copper mines on Butte Hill kept getting bigger and richer. Such ore bodies could stand any amount of champagne and bad management and political warfare. But there was one weak point in the copper-riveted armor of Anaconda. An enemy from within its own ranks found it. In the midst of the political war with Clark, Daly had his back to the wall in a still more desperate fight to keep from losing all the ore bodies he had developed.

F. Augustus Heinze was the traitor who found the vulnerable spot. At his name the Anaconda magnates still become incoherent with profane rage. They had found the richest mines in the world, and now a young upstart was trying to take them away again. And Fate had outdone herself to make Heinze a worthy hero and villain of the Butte comedy. Jew, Lutheran and Yankee ancestors gave him all their shrewdness. No scruples hampered his progress. His tall, handsome figure was the center of attraction wherever he went. A fine education put him at ease in the most cultivated Butte society, and mining-camp society can be polished to the hilt when it wants to be. In the dance halls and the corner saloons it was just the same. At a smile from Heinze the gamblers and the painted

F. Augustus Heinze twisted the apex laws to his own advantage.

ladies joined the copper magnates and the social leaders in adulation. Everybody was for him. A few months after he came from the Columbia School of Mines in '89 and started to work for the Boston and Montana, the highest positions the big company could offer seemed open to his ambition.

Heinze earned the hundred dollars a month the Boston and Montana gave him for running survey lines down in the drifts under Butte Hill. But the company would have been much better off if it had paid him a hundred thousand a month to stay away. The work his eyes and hands did for the B. and M. did not prevent his mind from always working for Heinze. It was not long before he saw the great chance that the highly paid executives had overlooked. There never had been such a glorious lot of easy money waiting to be picked. The only trouble was that he would have to get a good stake to start with. The new copper magnates would not give up without a struggle, and lawsuits cost money. So he kept his mouth shut until some lesser graft should pave the way for the great one.

The chance came sooner than he could have hoped. His grandmother died two years after he came to Butte and left him $50,000. This was better than a hundred a month, so he promptly quit his job and left Montana—but did not forget it. A few months of study and play in Germany celebrated his freedom. Then he came back to New York and took a position as assistant editor of the *Engineering and Mining Journal*. He needed rich friends to back his great venture, and this was the way to get them. The *Mining Journal* gave him a chance to meet wealthy New Yorkers who dabbled in mines, and his winning personality did the rest. In two years he had raised money enough to form the Montana Ore Purchasing Company and to build a smelter to treat ore from the smaller Butte mines.

The smelter was only a mask for his real operations, but it gave him the foothold he needed in Butte. And he made it pay, even if it was not the project he had at heart. Soon he was leasing and buying mines that could be picked up cheaply. His ancestry made him a good dealer, as the trusting miners soon found out. Jim Murray was foolish enough to give him a lease on the rich Estella claim on condition that returns from all ore running under 10 per cent would go to Heinze and those from ore over 10 per cent would be divided between Murray and Heinze. Both of them knew that the average grade of Estella ore was 12 per cent. But Heinze mined barren rock from the walls of his vein and mixed it with the ore. The result was that the material smelted assayed 9 per cent. Of course this gave him all the profits. You can't beat such business acumen – or ethics. Other deals were just as successful. From all hands the money rolled in, and in 1895 he was ready to go after the big game.

The law of the apex was at the bottom of the scheme. In the 1860s, when mining in America was young, Congress got all excited about some poor prospectors who were being robbed of the proceeds of their discoveries. The first mining laws had allowed the one who found an outcrop of ore to locate a piece of ground of a specified size and to mine all the ore within the vertical boundaries of the claim. This was all right if the ore went down steeply from the place where he found it. But sometimes he started to dig and found that the ore went off flatly to one side, like a sheet of cardboard with one edge a little higher than the other. Then the discoverer soon came to his sideline as he followed the vein down its flat incline. Anyone who was wise or lucky enough to locate the ground next door would get the continuation of the vein in depth, which was often the best part of it. The one who made the discovery was out of luck.

Congress in its all-seeing wisdom soon ended this injustice. The mining laws of 1866 and 1872 provided that the one who found a vein could locate a claim fifteen hundred feet long along the vein and six hundred feet wide. But his rights were not limited by the vertical boundaries of the claim. He could follow any vein that came to the surface or "apexed" in his claim just as far as he wanted to down the dip, provided he stayed within the projections of vertical planes passed through his parallel end lines. The "extralateral rights" thus given allowed him to mine far beyond his sidelines, under ground that belonged to his neighbors. So the discoverer was assured the fruits of his labor.

This was all right as long as the ore was in a simple straight vein like a sheet of cardboard. But nature does not work that way. A lot of veins are crooked, and vertical endline planes at right angles to them in one place cross those at right angles to them in another place. Here is a good chance for a fight. Then sometimes two veins start far apart and unite a few hundred feet underground to make one vein. Which of the claims on the two surface outcrops owns the vein below the junction? Often movement of the rocks has torn apart or "faulted" a vein, leaving one part of it many feet or hundreds of feet from the other. Who owns the part below the fault? Or maybe the ore does not come in flat sheets or veins at all, but in irregular lenses with outcrops too big to be included in one claim. Hundreds of other doubtful cases are caused by the infinite complexity of ore occurrence. Congress tried to straighten out the mess by saying the locator of the earliest claim could follow down his vein with all its "dips, spurs, and angles," regardless of conflicts with rights of claims of later date. It made the trouble almost worse than before, for no one knew just what a dip or a spur or an angle was. Lawyers and mining experts grew rich trying to settle the arguments in the courts, and the poor miner had no idea which end he was standing on.

In Butte in the early days the veins were steep and seemed fairly simple. Veins stayed inside the sidelines for many hundred feet below the surface. There was not much apex trouble. But after a decade or two the mines were approaching half a mile in vertical depth. Heinze saw that the increasing depth changed the apex situation completely. On the surface there was a network of minor veins of all sizes, joining or crossing the great major veins as well as one another. Often one vein would be shoved out of place by a fault that was itself a vein, or the fault had so little ore in it that it was anyone's guess whether it was a real vein or just a plane of displacement that had a few chunks of ore dragged along it from veins that it offset. As long as the network stayed within the vertical sidelines there was not much trouble. But a couple of thousand feet below the surface so many veins had joined or crossed that no one could be absolutely sure which was which. It was like a complex picture-puzzle with only occasional edges of the different jig-saw pieces exposed by the underground workings. A good lawyer and experts with the right sort of imagination could make out a convincing case wherever it was needed. And a friendly judge could always find a good reason for deciding the right way in the conflicting maze of testimony.

From his work in the Boston and Montana Heinze knew that a lot of branches came in to the big ore body. For a few thousand dollars he could buy the claims on which some of the branches outcropped. Then it would be up to the courts to say who owned the great ore bodies below the junctions. And Heinze had studied his fellow men enough to know that judges are human and have to eat. Here was his great chance, and he was at last ready to cash in on it.

Down on the lower slope of Butte Hill was a claim called the Rarus. There was a little rich ore in it, but the ore was hard to follow. It was no wonder, for years later geologists found that this ore was dragged in along a fault that had offset some of the biggest Butte veins many hundred feet. As they followed down the fault the Rarus owners ran into rich ore running down below the fault plane. This ore had no direct connection with veins that were being mined above the fault. A complex structure like this looked good, and in 1895 Heinze bought the Rarus for $400,000. He was just twenty-six years old then and spoiling for a fight.

He soon got it. Next to the Rarus was the rich mine of the Boston and Montana Company. Heinze had surveyed these ore bodies for B. and M. a few years before, and knew all about them. Now he could use the knowledge. A few months after he took over the Rarus, the B. and M. engineers heard rumors that he had crossed the sideline and was mining their ore. Heinze would not let them go underground, but rough

The Anaconda Mine for a long time was the greatest in Butte and a center of the "apex" warfare. Below, one of the earliest compressed-air drills works on the 1700-foot level in the Anaconda.

Montana Ore Purchasing Co.'s smelter, above, was at the east end of the Butte district. Below, the Rarus, Heinze's first mine, was on the east slope of the Hill.

surveys by miners they hired as spies indicated that the rumors were correct. They still looked on Heinze as one of the family, so they asked him to explain the trespass. He was very polite and regretful about it. The geology was complicated, and maybe there might be some grounds for uncertainty. He felt sure he was right, but to avoid hard feelings he would pay the B. and M. $250,000 for a quit-claim deed to the twenty or thirty million dollars' worth of ore. Colonel Bigelow, the courtly president of Boston and Montana, came as near apoplexy as his Boston training would let him. He would show this faithless young upstart that he could not trifle with the destiny of the emperors of copper. So he brought suit to enjoin Heinze from mining any more ore in the Rarus and consigned him to the nethermost depths of oblivion.

Before the suit came to trial the Amalgamated Copper Company had absorbed the Boston and Montana together with Anaconda and most of the other great Butte mines. This was a little more than Heinze had bargained for, but he jauntily set forth to battle with the trust that was to dominate the copper industry of the world, backed by all the power and ruthlessness of Standard Oil. It was a big contract. Heinze agreed with Bob Fitzsimmons that the bigger they are the harder they fall, and he made it the fight of a century.

The lawyers were the shock troops in this copper war. Amalgamated pinned its faith in William Scallon, a brilliant and honorable lawyer but no match in trickery or politics for the versatile Heinze. As attorney and managing director under Daly, Scallon controlled the destiny of the copper trust. Heinze was luckier in keeping a lot of his legal fees in the family. His brother, A. P. Heinze, gave up a good New York practice to study all the intricacies of mining law. His job was to find pretexts for claiming the ore Amalgamated was mining, and for tying up the big company by injunction. He found lots of them. Heinze and his Montana Ore Purchasing Company brought suits so fast that there was no chance to try them or to find what they really meant. A third Butte Superior Court judge and three special Masters to help the Montana Supreme Court could not stem the flood. At one time there were more than a hundred cases pending. The injunction that Bigelow had brought to drive Heinze out of Butte had such a prolific brood of pestiferous offspring that for a time it looked as though the mighty Amalgamated Copper might be overwhelmed by the sheer number of them. They were like seven-year locusts, only these locusts kept multiplying throughout the entire seven years the fight lasted.

Heinze knew that half the game lay in winning a friendly judge. This meant politics, and getting votes was easy for Heinze. All he had to do was to get up in front of a crowd of miners and denounce the iniquities of the infamous copper trust, with the Standard Oil yoke around its neck. Anyone would believe so likable and convincing a speaker – especially when he promised them shorter hours and higher wages. To avoid any possible defeat, he found an ally who was just as anxious to beat Daly as Heinze himself was. W. A. Clark had by this time won his appointment as senator, but did not dare take his seat until the legislature re-elected him. As long as he was "vindicated" he did not care what happened to Amalgamated or to Butte or to Silver Bow County. Daly's two enemies therefore formed an alliance. Clark helped Heinze with unlimited money for the local campaigns, and Heinze helped Clark with impassioned speeches to elect friendly members of the legislature.

This political combination was too strong for even Daly to resist. He was getting old now and sick, and death soon ended the fight for him. Scallon was far too weak to cope with the strangely assorted allies. Heinze won every local election during the seven-year fight. His judges, Clancy and Harney, decided every case that came to the Superior Court in Heinze's favor, and Clark won his million-dollar vindication.

Judge Clancy was the worst thorn in the side of Amalgamated. Until Heinze's money brought him to power, he was just an ignorant saloon loafer, picking up an uncertain living as a shyster lawyer. Now he was Superior Court judge of the richest county in the West. Dozens of suits involving tens of millions of dollars were tried before him. The geological structures involved were so intricate, and the expert testimony so technical and conflicting that the most brilliant and hard-working judge would have hardly been able to understand the issues. The complexity of the suits did not bother Clancy. While Winchell and Sales and the other Amalgamated witnesses were on the stand, illustrating their testimony by maps and models of the mine that had required months of hard work to prepare, Clancy sat with his back half turned and his feet on the window ledge, stroking his long gray beard and gazing vacantly at the blue of the distant mountains. At exactly spaced intervals he turned his head a fraction of an inch and with a loud splash added to the pool of tobacco juice in the flaring brass cuspidor. He didn't have to worry about what the witness said. His verdict had already been written by Heinze's attorneys, and Heinze would see that he was re-elected by a big majority.

It was a discouraging job testifying for Amalgamated before Clancy. But the witnesses knew the cases would be appealed. They had to give all their evidence even if Clancy did not listen to a word of it.

If it had not been for the higher courts, Amalgamated would have been ruined. The apex law threatened its most valuable mines. Even when there was no chance of actually taking the ore bodies away from the company, the Superior Court issued injunctions against mining the contested ore until ownership was settled. This involved proof of "identity and continuity" of the veins from the surface down. Both Heinze and Amalgamated had to run thousands of feet of drifts and raises on the ore to prove that a vein half a mile underground was the same as an insignificant outcrop on the surface. Meanwhile trial of the suit was postponed and one or both sides were enjoined against mining. The dozens of injunctions so reduced the production and profit of the big company that dividends were cut and the price of Amalgamated stock dropped many dollars a share. Rogers and his Standard Oil associates in New York swore vengeance, but for years they were helpless.

Most of the suits and counter-suits were just intended to befog the issues and clutter up the courts. Three or four cases were really vital. The Copper Trust, Michael Davitt, and Minnie Healy cases were the most serious, and the fight centered on them.

The Copper Trust claim was a little wedge of ground seventy-five feet long and ten feet wide at the base. Heinze bought it for a song. After legal and geological preparation he brought suit on the ground that the ore bodies in the Anaconda, St. Lawrence and Neversweat mines – the cream of the Amalgamated property – apexed in this little fraction of a claim. It sounded ridiculous. But Heinze drew beautiful geological sections that showed a great vein running straight from an outcrop in the Copper Trust to the ore bodies he coveted. The sections bore absolutely no relation to the true geology, but Judge Clancy accepted them as correct and issued an injunction stopping all production from the three great mines until the apex could be proved by slow, underground work.

The situation looked hopeless, as legally there was no recourse. Luckily Heinze had not figured on the effect of the injunction on the thousands of men who would be thrown out of work by shutting down the mines. As the men came off shift the afternoon of the injunction the bosses told them it was up to them to save their jobs. They acted quickly. That same night three thousand angry miners marched down Butte Hill. Shouts of "Hang Clancy" rang out on all sides. The police were powerless – and kept out of the way. Judge Clancy hid out for a couple of hours, but he had no desire to be a dead hero. At midnight he dissolved the injunction, and the shift went to work the next morning as usual. The injunction had lasted just fourteen hours. The higher courts ruled after several years that the Copper Trust had no right to the ore.

The Minnie Healy case was a much tougher one for Amalgamated. This claim was in the heart of the richest area on

The Leonard Mine, above, and the adjoining Minnie Healy, below, set the scene for another long-standing legal fight.

Butte Hill, but for some reason ore was hard to find in the Minnie Healy. Daly had taken an option from Miles Finlen and spent over $50,000 in development without any success. This disgusted him so that he threw up the option. Heinze was waiting eagerly for this chance. The Minnie Healy was next to the big Leonard mine of Boston and Montana, which was soon to be absorbed by Amalgamated. A barren branch vein made it easy to claim that the Leonard ore apexed in the Minnie Healy. So as soon as Daly quit work Heinze took an option from Finlen, on condition that Finlen keep the deal secret until the suit was brought. The papers were drawn, and Finlen agreed verbally. But he wanted some minor changes. As there was not time to copy the papers before he left on a short trip to New York he did not sign, but promised to do so on his return.

Heinze started work to prove his apex. He had no hope of finding ore. All he wanted to do was to keep his underground drifts and raises in some structure that he could claim was a vein. Suddenly a raise broke into high-grade copper sulphide. The unexpected ore body soon proved to be one of the richest in Butte.

Daly could not stand for this. Heinze not only had a claim that threatened the Leonard, but he had had the insolence to pick this rich plum out from under Daly's very nose. In a few days lawsuits flew right and left. Heinze claimed the Leonard apex, and Daly induced Finlen to repudiate his verbal contract and to sell the Minnie Healy to Amalgamated. This time Amalgamated kept the case out of Judge Clancy's court. The other Superior Court judge, E. W. Harney, was an educated lawyer and had been a good judge when he was not drunk. The decision was delayed until 1901; then Harney found that the Minnie Healy belonged to Heinze. Through Mrs. Ada Brackett, a public stenographer with whom the judge was on suspiciously intimate terms, detectives for the company got correspondence that proved Heinze had bribed Harney. On the other side Harney himself testified that an Amalgamated agent had offered him $125,000 for a favorable decision. The lower court ruled that the evidence was not conclusive enough to warrant impeachment, but the Supreme Court reversed the Minnie Healy decision and sent the case back to Judge Clancy for retrial. It was still pending at the end of the Heinze-Amalgamated fight. All through the long litigation Heinze kept possession. Whether the Minnie Healy was his or not, it kept his smelter going for six years.

The Michael Davitt case was even more bitterly fought. The Butte and Boston Copper Company was mining the rich Enargite and Windlass veins in this claim. Heinze owned the adjoining Rarus claim, and mined across the endline under the Michael Davitt. The Butte and Boston brought suit to keep him out. This was a jury case in the Federal Court. In the first trial in 1898 the judge instructed the jury to find for Butte and Boston, but it refused. The next trial was in Helena in 1900, after Amalgamated had taken over Butte and Boston. Heinze stirred up a wave of popular indignation against Standard Oil and Amalgamated, and the jury decided for him. Judge Knowles granted a third trial on the grounds of undue influence. It was delayed two years, with both sides enjoined from working the ore bodies.

Heinze was then desperate. The ore in his own mines was not big enough to pay the enormous litigation costs. He had to find some way to evade the injunction. He assigned his right to the ore to a new company — the Johnstown Mining Company. This, he said, was not bound by court order. So the Johnstown Company started to work again in the forbidden area.

Heinze's superintendent, Al Frank, had mined the Enargite ore out to within a few feet of his endline on the seventh, eighth, and ninth levels. In the "pillar" of unmined ore that he had to leave for safety next to the line, he ran raises between levels. Twenty feet below every level he started crosscuts from these raises, running out in the barren rock of the vein walls. The crosscuts turned like corkscrews, so that no one could tell without a survey just where they were. After fifty feet of apparently aimless wandering, they came back to the vein again inside the Michael Davitt, where Amalgamated had been forced by the injunction to leave a large block of beautiful ore. The court order allowed Heinze's engineers to enter and survey the Michael Davitt workings. With this information they could easily plan the drifts and stopes so that they would not break into workings of the enemy. Protected by this knowledge and by his corkscrew drifts, Frank sent in a large force of miners and started to tear out the ore as fast as he could.

The Amalgamated foremen could hear Heinze's blasts from their Berkeley Mine, which was next to the Michael Davitt. But for months they could not do anything about it.

At last Reno Sales, assistant geologist for Amalgamated, figured out a way to get into the Rarus. A connecting drift had been run between the Berkeley and the Rarus some years before. It was now blocked by a concrete door or bulkhead. Between shifts one February night in 1903 Sales went into the old drifts with McGee and Finnegan, superintendent and foreman of Boston and Montana, and with steel bars dug a hole around the bulkhead. Crawling through the hole they found another bulkhead blocking their way. They dug around this in turn. On the Rarus side of it there was an arm chair, and a telephone. Evidently Heinze had kept a watchman there to guard against invasion, but for some reason the watchman was off duty.

Sales and his companions slowly made their way toward the Michael Davitt endlines. They almost fell into "winzes," or deep holes left open in the bottom of the drift as traps. A fall of fifty or a hundred feet might result from a single careless step. By the flickering light of their candles they crept through the black deserted drifts. Every drop of water resounded in the stillness of the mine like a clap of thunder. Their hearts hammered like compressed-air drills. At last they breathed freely again when they realized that not another soul was in the mine. After wandering through many hundred feet of workings the rough survey that they made as they went along showed that they were at the endline. Their course was blocked by a wall of solid rock. There must be some other way into the Michael Davitt. Sales noticed that the ladders in a raise a few feet back from the face were badly worn. Many men must have used them. He climbed down the raise, and found the corkscrew crosscut into the wall. At the end of it was a great open stope, two hundred feet long and thirty feet wide. A hasty survey showed that the crosscut had come back to the Enargite vein in the Michael Davitt, and this was where the stolen ore was coming from. Their venture had succeeded. The three invaders hurried back to the Berkeley before the Rarus workmen came on shift and found that their secret had been discovered.

Amalgamated now had evidence that proved the trespass. It was another thing to stop it. The Federal Court issued an order allowing Horace Winchell, chief geologist for Amalgamated, to enter the Rarus with his assistants and survey the damage. They tried all the underground connections, but found them blocked by concrete. Then armed with their court order Jack Adams, the Amalgamated superintendent in charge of the "Firing Line," went with Winchell and Sales to the collar of the Rarus Shaft to try to get down. Nick Treloar, the old "Cousin Jack" foreman whom they had known for years, was sitting on a stick of timber near the shaft.

"It's a good-looking mine you've got here, Nick. Who owns it?" Adams asked.

"I don't know," replied Treloar.

"Doesn't it belong to Heinze?"

"I don't know," said Treloar again.

"Well, we've got a court order to let us down, and we want to go."

"I haven't got anything to do about it."

"Who in hell is boss around here if you aren't?"

"I don't know," Treloar once more replied.

The skip tender at the shaft said he couldn't let anyone down without an order from the boss. Who was the boss? He didn't know. The visitors brushed past him and climbed on the cage. They rang the signal for the hoisting engineer to let them down. The cage stayed still. The engineer could not move it unless the boss gave the order. Who was the boss? He didn't know.

For months the Amalgamated engineers tried their best to get underground. Heinze's well-trained employees balked every move. And meanwhile they continued to tear out the stolen ore.

At last the Federal Court ordered Heinze to appear in person at the shaft collar and let Adams and his men down in company with a United States marshal. Heinze evaded service of the order for a few days by climbing out of a back window of his office. Finally it was served, and he could not keep them out any longer. They went down the endline raises and through the corkscrew drifts and proved again that Heinze was violating the injunction and stealing Michael Davitt ore. The marshal ordered all work to stop. It did for a minute, but before the marshal was out of the mine Heinze was mining again as fast as ever.

In spite of this direct evidence, Heinze succeeded in securing a few weeks more of legal delay. He claimed that Judge Knowles was biased. Judge Beatty of the Idaho District Court was brought in. After hearing the evidence, the new judge imposed heavy fines on Heinze and his superintendents for evading the court orders. Rather than pay the fines, Heinze at last let Winchell and his assistants go down the Rarus to measure the stolen ore.

October had already come. Nine months had elapsed since Sales had made his first trip into the trespass stopes. Heinze had stolen more than a million dollars' worth of ore and had almost exhausted the ore body.

Even now Heinze would not give in. When the Amalgamated party went down the Rarus to make the final survey, they found straight crosscuts in place of the corkscrews at the endline. The enemy had walled up all the entrances to the Michael Davitt stopes and was apparently driving harmless prospecting drifts in barren rock. As fast as the Amalgamated party tore down a wall, it was built up again.

As there was no chance to get the evidence through the Rarus, the Court gave Amalgamated permission to drive a crosscut from the Pennsylvania Mine to a point forty feet below Heinze's stopes, and to raise up to them. The raise broke through on January 1, 1904.

Heinze's men were ready to repel the invasion. This was war, and court orders meant nothing a thousand feet underground. As the Amalgamated miners tried to climb up the raise, an avalanche of rocks drove them out again. Two miners were hit and badly bruised. Jack Adams and his men then lifted timber shields above them and tried it again. They had not counted on the air current that flowed from the Rarus down the raise and into the Pennsylvania. Heinze's miners burned old clothes and boots near the top of the raise and drove them out with the stench. When the fire died down Adams made another attempt. The Rarus crew then put dry slaked lime in the high-pressure compressed-air line and blew it down the raise. The cloud of dust almost choked Adams and his crew. Still they kept on trying. The Heinze army was desperate. As a last resort they set off charges of dynamite near the top of the raise. The poisonous fumes drifted down and knocked two Pennsylvania miners unconscious. Their companions found them and dragged them back to fresh air just in time to save their lives.

The Amalgamated miners at the foot of the raise heard a Rarus shift boss above give orders to set off a heavier charge of dynamite if there was another invasion. Their shift was over, and they were glad to let someone else take up the fight. As they went out to the shaft they told their successors on the next shift, Samuel Olsen and Fredolin Divel, to look out. The report was passed on to Jack Adams. He realized that he had the worst of it and must retreat to avoid having his miners killed and his whole mine made unendurable by the dust and fumes. He went to the bottom of the raise and ordered Olsen and Divel to keep out of the raise itself, but to build a wooden door in the drift near it to stop the air current. They worked fast, and soon had the door almost com-

The Pennsylvania and Neversweat mines, above and below respectively, which later became Anaconda properties, were among the great ones south and southwest of Butte Hill.

pleted. Just as they were ready to leave, a Rarus miner lowered a box of dynamite with a burning fuse down the raise. Olsen and Divel had no chance to get away. The dynamite exploded, blew the door over on them and killed them both instantly. A Pennsylvania shift boss who was coming in to inspect the work was overcome by the powder smoke and barely escaped with his life.

A coroner's jury found that the deaths were due to criminal intent of the Rarus miners. There was no criminal prosecution, though feeling ran high. Heinze was powerful enough to prevent action by the local courts. The widow of one of the men who was killed finally got a $25,000 verdict for damages against Heinze. That was the only penalty.

The Michael Davitt case ended three months later with a decision by Judge Beatty fining Heinze $20,000 and his superintendents $1,000 each for violating the court order. The Judge called Heinze's claim that the Johnstown Mining Company was not bound by the order flagrant quibbling. Heinze paid the fines cheerfully. A fine of $22,000 was an easy way to get out of stealing a million dollars' worth of ore.

Meanwhile the underground war spread to the Horsetail ore bodies in the Minnie Healy area. In the Leonard mine of Amalgamated, north of the Minnie Healy, the east end of the great Anaconda vein turned to the south and fanned out like the hairs in a horse's tail until the rich ore was hundreds of feet wide. This ore crossed the Minnie Healy sideline at a depth of a thousand feet or more. Heinze claimed that one of the many branch stringers gave the Minnie Healy the apex of the whole Horsetail ore body. Judge Clancy had obligingly enjoined Amalgamated from doing any more mining until the apex was settled, and the case had tied up the Horsetail area for years.

Heinze had to have more ore to keep his smelter running, or he was lost. Here was his chance. The Minnie Healy itself might not be his, and the apex remained to be proved, but he had possession of the Minnie Healy and could easily reach the ore. Once it was mined, Amalgamated could whistle for its damages. He crossed the sideline and started to mine the ore.

Miners in the Leonard heard the blasting and realized that Heinze was trespassing again. Reno Sales hunted through the labyrinth of old workings until he found an open connection to the Minnie Healy. Following a crosscut to the Leonard sideline he discovered a great untimbered hole two hundred feet long, fifty feet wide and twenty feet high. Heinze was mining the ore out so fast that he did not even stop to protect the workings from caving. As Sales entered the stope four Heinze miners were at work mining the high grade. When they caught sight of his light, they fled away into the maze of workings. He followed. Everywhere the ore thieves disappeared ahead of him. There must have been dozens of them. But only a twinkle of light from a distant candle and the acrid smell of stale powder smoke proved that the mine was not deserted. Sales came back to the big hole where he had found the four miners. In the ten minutes he had been away the broken ore had disappeared. It was as though a band of gnomes had spirited it away under his very nose.

Amalgamated had to act quickly to save the rest of the ore body. Jack Adams gave leases on the ground covered by the injunction to picked miners who were good fighters. He claimed that the lessees were not subject to the court order — as Heinze had done with the Johnstown Mining Company. The men who were making ten dollars a day in a good lease would take risks on their own account that he could not ask them to take. The miners were for the most part Irishmen who loved a good fight on general principles. An ordinary street fight with fists or clubs was all right as far as it went. But an underground war to the finish, with unknown dangers lurking around every turn in the inky black workings, made their blood curdle with happy excitement. The Butte miners of thirty years ago are old men now, but whenever they get together over a glass of beer they tell of the winter of the Heinze war as the great adventure of their lives.

The two companies rushed new crosscuts to the ore body. Adams was nearer and usually got there first. He could hear the Minnie Healy miners drilling as they approached the ore. His levels were a few feet lower than those in the Minnie Healy. With this advantage he could wait till a Heinze crosscut was just over one from the Leonard, and then set off a heavy charge of powder and wreck it. A warning notice before the blast gave the enemy miners time to get out of the way.

Sometimes an invading drift would get far into enemy territory before it was discovered. Adams gave an order one day to start a raise from a drift that he thought was fifty feet from any Heinze workings. That afternoon Reno Sales went to look at it, and found the Italian miner in a frenzy of terror. The miner had started a drill hole, and it broke into what he thought was a crevice. He withdrew the drill and stuck a bar up in the hole to find how big the crevice was. Suddenly the bar was snatched from his hands, flew up in the hole, and disappeared. He thought it must be the devil who had done it. Sales had a hard time calming him down and making him realize that he had drilled into a Heinze drift. A miner there had seen the bar and had yanked it up as a joke.

The blasted crosscuts delayed Heinze, but others soon reached the ore. Both companies started to mine at top speed, side by side. Every man had to protect his own little section of the vein as best he could. The one who was handiest with his fists or with a good heavy pick handle got the most ore and the biggest pay check. As the fighting became hotter, a resourceful miner invented a hand grenade. He put half a stick of dynamite, with a cap and a short fuse, in a tin can, lighted the fuse, and tossed the can just close enough to his opponent to put out his light and to scare him half to death without actually killing him. Both sides enthusiastically adopted this new weapon. Bombs were soon popping all over the Horsetail area. It took a real man to mine and to fight on a rough pile of broken ore, with the black walls of the stope half lighted by the flame of a candle that flickered despondently in an atmosphere saturated with steaming moisture and with the sour smell of sulphide dust and sweat, plunged in darkness now and then by a bomb that almost knocked him down. Often a misstep meant a fall of fifty or a hundred feet down the ore chute. Broken heads were common, but miner's luck kept anyone from being killed in the Horsetail war.

Month by month the fighting spread and became more violent. The Amalgamated officials knew that as soon as they were forced out of an area by a court injunction Heinze would be close on their heels ready to mine the ore. In one place a decision by Judge Clancy would put at Heinze's mercy a section of one of the richest veins in Butte nearly a thousand feet long. By the time they could get the appeal decided by a higher court Al Frank would have the ore all mined out. They weren't going to stand for this. Just before the decision was due they placed dozens of heavy charges of dynamite at all the entrances to the stopes. Of course Judge Clancy awarded the ore to Heinze. A code message, "Stormy weather," was rushed underground to the Amalgamated bosses. They hurried to the "firing line" and warned the Heinze miners to get out at once or get blasted. Half an hour after the decision the blasts went off. The whole area was a hopeless wreck: stopes caved, crosscuts filled with rock and machines destroyed. Heinze had no chance to mine that ore before the Supreme Court reversed the decision and gave it back to Amalgamated.

Early in 1904 the fighting in the Horsetail ore bodies changed from guerilla warfare to an organized campaign backed by all the resources of the mines. Heinze was as usual barely staving off his bank overdrafts. He had to get more ore. In the Minnie Healy as in the Rarus the air currents favored him. So he told Al Frank to resort to the same tactics he had used in the Rarus-Michael Davitt battle. Frank soon made life unendurable in the Leonard crosscuts by burning dynamite and rubbish near the underground connections. Where the Amalgamated miners were within reach, he alternately blew steam and dry powdered lime at them through the air pipes. Slowly he drove the enemy back until most of the Horsetail area was in his hands.

Jack Adams found a new weapon with which to resist the invasion. In the Leonard shaft there was an eight-inch pipe or "column" that carried the water under pressure of hundreds of pounds per square inch from the underground pumps to the surface. To help in fighting underground fires, this shaft column was connected with the main Butte water supply. On every level branch pipes led to convenient places in the drifts. Reels of fire hose were mounted on trucks ready for emergencies. This was certainly emergency enough. Adams ordered the fire hose coupled to the high-pressure pipe line and strung it along the drifts to the entrance to Heinze's stopes. In a few moments the terrific stream drove all the Minnie Healy miners back toward their shaft. Adams had won the second phase of the big battle.

Frank replied with a barrage of smoke, steam and lime that made it almost impossible to breathe in the Leonard. His weapons had the longest range, and he felt sure of victory.

Adams was back to his last trench. Only a desperate assault could save Amalgamated from the loss of its most valuable ore bodies. The plans had been laid long before but Adams had hesitated to use them. Now the time had come. Al Frank had beaten Adams in the Michael Davitt and Adams must win at any cost in the Horsetail stopes.

The Minnie Healy shaft was within a few feet of the sideline of the claim, next to the Leonard. Working between shifts, when the Minnie Healy miners could have no inkling of what was happening, Adams had driven a Leonard crosscut to within four feet of the Minnie Healy shaft. In the crosscut he laid a four-inch pipe, leading from the Leonard water column. A few minutes' silent work with a diamond drill cut through the four-foot wall of rock to the Minnie Healy shaft. The Leonard pipe was run through the hole, ending behind the timbers where no Heinze shaft men would notice it. With the turn of a valve all the water from the Amalgamated mines and the City of Butte would roar through the pipe into the Minnie Healy.

Adams was ready for the *coup de grâce*. He gave Al Frank fifteen minutes in which to get all his men away from the shaft. Frank knew something desperate was about to happen and ordered his miners to the surface or back in the drifts. The moment the fifteen minutes were up, an Amalgamated boss opened the big valve. Under hundreds of pounds of pressure the water rushed through the big pipe and poured into the Minnie Healy shaft. It tore out timbers and loose rocks and in a few moments completely wrecked Heinze's only entrance to the mine.

Beaten underground, Heinze made one desperate attempt to drive out the enemy by an appeal to the mob. With his own hands he blew long blasts on the Minnie Healy whistle. Only a terrible accident could cause this shrill warning in the middle of the shift. Miners who were off shift and their families and hangers-on from the saloons hurried to the shaft collar. Rumors began to fly. Someone whispered that Amalgamated had flooded the shaft without warning and drowned the Minnie Healy miners like rats. The whispers grew to angry roars from fifteen hundred Heinze sympathizers. Someone yelled "Lynch them," and the crowd took up the cry with shouts of rage. With a rush the mob started toward the Amalgamated office, where Scallon and his superintendents were waiting. It looked like the last shift for them.

Wallace Corbett, superintendent of the Leonard, saved the day. With a loaded rifle he went to the Heinze mine office. Heinze's superintendents must tell the mob the truth, he said, or they would never live to see Scallon hanged. Unwillingly they hurried to the head of the lynching party and told the leaders that the rumors were false. The mine had been flooded, but they had been given notice and no one was hurt. They were just in time. The mob hesitated, but finally believed them and went silently home.

This was the end of the underground war. Both sides had enough of it. For a while they had enjoyed the excitement, but mob violence and wholesale murder were a little more than they had bargained for. As Al Frank put it, "It's too savage for us." The bank accounts of Heinze and of the "Standard Oil crowd" did not mean that much to them. A few of the leaders talked it over the evening of the near-lynching, and next morning the superintendents and engineers on both sides met in the Silver Bow Club. It was a good place for the "peace conference," as they had met there many times for poker games and drinking parties before the war began. They were naturally friends, not enemies. With their eyes opened by the narrow escapes of the afternoon before, they soon came to an unwritten agreement. Both sides agreed to abide by injunctions and to let the courts decide the differences between the companies.

The fight in the courts continued for two years more. Clancy and Harney decided all cases in favor of Heinze in the Superior Courts, and the Montana Supreme Court and the Federal Courts usually gave the verdict to Amalgamated. It was a hopeless tangle. For every case that finished the slow round of the higher courts three or four new ones sprang up. Heinze saw every dollar he took out of the ground go to the lawyers and experts. And Amalgamated lost the profits of its richest subsidiary when Heinze secured an injunction that forbade the payment of Boston and Montana dividends to Amalgamated on the ground that the consolidation had been illegal. There was no end in sight but bankruptcy.

The Standard Oil and Amalgamated directors learned after many years that their arbitrary "To hell with the public" policy did not pay. Goodwill was necessary to permanent success in business. They made John D. Ryan managing director in Butte to win over the public opinion that had been inflamed by Heinze against the Octopus of Oil and Copper, and to end the constant fighting. To help end the Heinze war Mr. Ryan brought in Thomas F. Cole, a master negotiator who had successfully assembled the best of the iron mines in the Lake Superior region for the Steel Corporation. Heinze came high, but they finally bought him out. In February, 1906, the

Butte Coalition Mining Company, a new subsidiary of Amalgamated, took over all the Butte properties of Heinze's intricate companies.

Heinze was out at last, and Amalgamated could count the cost of the war. At the start of the fighting Henry H. Rogers had said in one of his flashing bursts of rage, "We'll drive Heinze out of Montana if it takes $10,000,000." It would have been a bargain at that. The litigation cost Amalgamated a million dollars a year. And at the end Butte Coalition paid Heinze $14,000,000 in cash and a large stock interest.

With the settlement eighty lawsuits involving $100,000,000 were dismissed. The Montana courts could attend to their regular business for the first time in six years. And the superintendents and engineers could at last go back to mining, instead of working up evidence for lawsuits.

The litigation seemed to end in a victory for Heinze. Really Amalgamated got all the best of it. Heinze was through. With his $14,000,000 he tried to start a rival copper trust called the United Copper Company and a string of New York banks. But his lucky star had deserted him. Most of his mines outside of Butte proved worthless, and Standard Oil opposition soon wrecked the banks. For the last few years of his life Heinze was just a promoter of doubtful reputation, living on his past glory.

For Amalgamated the years of litigation proved to be worth far more than they cost. Unexpectedly they brought forth a new science, and new ore bodies that no one had dreamed of. Before the Butte legal war started, there was hardly any such thing as mining geology. The Geological Survey got out a lot of elaborate papers, but "practical mining men" laughed at them. The mine foreman and superintendents ran their exploration drifts where their "nose for ore" called them. Their idea of ore occurrence was summed up in the Cornish miner's proverb: "Where she be, there she be." And as for geology, one old Butte superintendent summed it up perfectly when he told a director: "There are just three kinds of rocks – ore, waste and water."

So geology was in ill repute in the early days. A good old Irish foreman would hardly allow "one of them yellow-legged sons of —" underground. Anyone who suggested that the geologist could find twice as much ore as the foreman did would have been thrown out on his neck.

The apex litigation changed all that. To convince a judge or jury who knew nothing of mining that a vein two thousand feet underground was the same as a barren "leached" outcrop on the surface required clear thinking and a flow of words far beyond anything a self-made mine boss could muster. Regretfully the companies hired engineers and geologists to go on the witness stand as experts. At first the experts knew little more than the miners, for the study of ore occurrence was still young. Gradually they learned more about it. The picture puzzle of crisscross veins and faults began to assume an intelligible pattern. The geologists planned drifts and raises to prove their theories to the court. The foremen did not want to spend money on this development, but they had to. And to their surprise many of the "litigation drifts" found rich ore they had not suspected.

Amalgamated was lucky in securing an ambitious geologist named Horace Winchell to take charge of its litigation work. Reno Sales was his young assistant. Together they worked out a plan of accurate geological mapping, supplemented by models of the mine workings. Gradually vein systems of baffling complexity became clear. Winchell and Sales generally convinced the upper courts that they were right. And the development work they planned in the unsuspected cross veins and in blocks that had been displaced by the countless faults found ore that was worth many times the cost of the litigation.

After a few years even the crustiest old foremen were calling on the scorned scientists for help. Mr. Ryan gave Winchell and Sales complete charge over all development work. For the first time geology had really become useful. The results of intelligent imagination added to the most painstaking record of observation were so unanswerable that one by one other mining companies turned their exploration over to scientifically trained geologists. The Butte litigation had started a great new branch of applied science.

With the end of the Heinze war, romance at Butte was dead. Efficiency came to take its place. Butte was not so exciting to read about, but it was a lot more satisfactory to the stockholders and to the miners and to everyone else. Wages were high, and working conditions became a lot better. When the smelters moved to Anaconda and Great Falls, it was possible to breathe in Butte without choking on the sulphur smoke. Occasional blades of grass began to show their heads in the bare front yards. The movie houses took away some of the patronage from the saloons and dance halls. Butte became civilized and dull.

And the dollars rolled in to the company treasury at an incredible rate. For fifteen years the mines made good the chamber of commerce boast that Butte was the Richest Hill on Earth. A second billion dollars' worth of copper, gold and silver was added to the billion that had been produced in the first forty years. The mines weren't quite so rich as they had been in the bonanza days, but more economical methods of mining and ore treatment more than made up for that. Butte was on the top of the wave.

It was too good to last. In copper mines as in all human affairs youth must be served. The mines are now 4,000 feet deep. There is still a lot of rich ore left. But the deep shafts and the hundreds of miles of underground workings add relentlessly to the difficulty of mining. Engineers have performed miracles in increasing efficiency. In the long run they know it is a hopeless fight. Twenty years ago Butte was the queen of all the copper districts. Twenty years in the future she will be only a half-forgotten dowager, sadly watching the success of the younger mines that drove her from the throne.

The wonderful copper properties that Mr. Ryan and his associates developed and bought in newer districts assure the Anaconda company leadership in the copper industry for a generation or more. These younger sisters will carry Butte along with them – paying in good years and often losing nearly as much in bad years. They can only postpone the end. One by one the great hoists will cease to turn, and the buildings will crumble into ruin.

In ancient times a copper mine was great for scores of centuries. With the mad pace of modern industry the path from wilderness to wilderness covers hardly a lifetime. Before we realize it the shades of Clark and Daly and Heinze will stalk through the deserted streets of a ghost city, and the Richest Hill on Earth will be only a memory.

AUTHOR'S NOTE: The most vigorous protest against *Romantic Copper* came from Alfred Frank, formerly superintendent for the Heinze mines in Butte. Frank's long letter to me was friendly, but he thought I had been unfair to Heinze and his associates. As most of my old friends were on the other side of the fence, the criticism may have some truth. Frank emphasizes the fact that there were "traitorous" actions on both sides. He says that the decisions of Judge Knowles against Heinze were just as "unfailing" as those of Judges Harney and Clancy for him. Mr. Frank comments at length on the apex lawsuits at Butte. He admits that he and the other Heinze chiefs were "fined for evading court orders," but complains that Horace Winchell and others on the Amalgamated side were not fined for violations in other cases.

Regarding the Minnie Healy, Frank says that much of the ore was taken by Heinze from within the sidelines of this claim. He says the Supreme Court of Montana finally gave the title to Heinze. About the so-called lynch threats by Heinze sympathizers, Mr. Frank wrote: "I myself did go to the Leonard main office, where I saw Wallace Corbett, who was armed. I demanded that the water be shut off, and it was *shut off*. There was a large gathering of men, sympathetic to our side, but to say that there were threats of lynching Mr. Scallon is too utterly ridiculous." Mr. Frank concludes: "I am proud of the fact that through it all and since that time I was generally able to maintain the most pleasant personal relations with our opponents."

I am sorry Al Frank is no longer alive and cannot see that I am giving this publicity to his side of the Butte story. At the time, I suggested to Frank that he and Reno Sales of Anaconda, who had a high regard for each other in spite of their differences, should write a joint account of the Butte war. This interesting chapter in mining history was never written, although in 1964 Reno Sales published his recollections in a book, *Underground Warfare at Butte*.

Now that 1973 has come, I am glad to admit that my conclusion of forty years ago to the chapter on the Richest Hill on Earth was far too pessimistic. The chapter in Part II of this book on the great low-grade deposits in the United States will show that the lean material around and between the Butte veins is being mined with a profit nearly as great as that from the veins themselves.

5

Luck and the Copper Queen —The Story of Bisbee

DID James Douglas make the Copper Queen or did the Copper Queen make James Douglas? History gives the credit to Douglas and says that his genius was responsible for the mine and through it lifted the Phelps Dodge Company to the pinnacle of success and respectability. But history never looks too critically at the early days of its heroes. In the case of the Copper Queen and James Douglas, Fate seems to have done the dealing, as usual. She ruined some of the players and bestowed fame and fortune on others. The combination of James Douglas with the Copper Queen mine proved to be the winning hand in spite of the fact that neither one of them had taken any tricks worth mentioning before they came together.

When he first visited the new Arizona copper camp of Bisbee in the fall of 1880, James Douglas had shown no indication of becoming the king pin of the mining industry or of any other profession. He had lived forty-three years without really deciding what profession suited him. Most of the years he had spent acquiring a lot of education in Canada and abroad. As a result of it he was licensed to preach in the Presbyterian church and to practice medicine in the Province of Quebec. He tried them both. For a year or two he was assistant minister of St. Andrew's Church in Quebec. Then for a while he managed the Canadian asylum in which his father had been one of the first to introduce the modern, humane care of the insane. Maybe his association with crazy men

Dr. James Douglas after several failures became the leading mining engineer of his time, making Phelps Dodge a great company. *(—Engineering Societies Library)*

helped him later on when he was dealing with Arizona prospectors. But he soon had enough of it as a regular occupation. Teaching chemistry in a small Canadian college filled in the next few years. About this time his father lost most of his modest fortune in the Harvey Hill copper mine in southern Quebec. James Douglas turned to the study of copper metallurgy to try to get it back again.

This research resulted in the Hunt and Douglas copper process. Dr. Douglas persuaded New York friends to build a plant to try out this method of recovery in Phoenixville, Pennsylvania. The process was only half successful. When the plant burned down a year or two after it started, no one would pay for a new one and Dr. Douglas was out of a job again.

So far, James Douglas was a highly cultivated Canadian gentleman, and that was all. He might almost have qualified as a drifter and a failure, except for the fact that everyone who met him liked and trusted him. His friends were sure he ought to amount to something some time. Some of them were even willing to accept his judgment on a mine, in spite of his lack of experience. Their confidence gave him a chance to bring the hard-luck half of his career to a climax with one of the best advertised mistakes in mining history.

A small metal-dealing concern named Phelps Dodge and Company had received reports of a rich copper prospect called the Verde, in the Black Range of northern Arizona. They sent Dr. Douglas out to examine it, early in 1880. He found a little showing of rich copper ore three or four hundred miles across the desert and mountain from the nearest railroad. Without hesitation he turned the mine down, saying that even if it was good it was too remote to have any possible value. This report killed Jerome and the United Verde Mine for nearly ten years. Then Senator Clark came down from Butte and made $60,000,000 profit out of it. Ever since then prospectors all through the Southwest have claimed that if a distinguished engineer condemns their mine it must be a good one, because Dr. Douglas turned down the United Verde.

On his way home from Jerome, Dr. Douglas stopped at Bisbee. Up to that time Bisbee had been even less successful than Dr. Douglas. Three years before it had been only a nameless cañon in the Mule Mountains. An army scout named John Dunn had ridden out from Fort Huachuca, across the San Pedro Valley, and was attracted by the red iron outcrops where Bisbee is now located. After a short time he found a little lead ore. As he was too busy to develop the showing himself, he grubstaked a worthless prospector named George Warren. Warren persuaded four other Tombstone miners to accompany him to the new camp. Warren found some lead and copper outcrops, and staked two or three claims. The other prospectors located the adjoining ground. They named the new camp the Warren Mining District.

George Warren soon went on one of his usual drunks. On the Fourth of July he bet the claim that afterward became the Copper Queen on a foot race and lost it. He sold the rest of his property for $925, with which he drank himself insane. After a few months in jail he recovered enough so that the sheriff let him out. He wandered across into Mexico. There he was thrown in jail again and peoned because he could not pay for the mescal he had drunk. Judge Berry of Tombstone took pity on him and paid the bills. Bleary-eyed and bewildered, poor old Warren came back to Bisbee to be the village bum. He lived for three or four years more, getting a little pension from the mines and doing odd jobs for his whisky money. That was all Fate gave the discoverer of the Warren District and the Copper Queen.

The lead outcrop that first attracted Dunn did not amount to much. The stringers of rich copper carbonate that Warren and the other prospectors found on the lower slopes of a big gray limestone hill west of the gulch looked better. Still Warren's successors did not think enough of it to do the necessary hundred dollars' worth of assessment work on the original "Mercey" claim. They let it lapse, and relocated it as the Copper Queen. The copper might become valuable some day, and it cost hardly anything to relocate the ground. At the worst, the showing might coax a few dollars out of some eastern sucker.

A shrewd Tucson merchant named Zeckendorff heard of the copper strike and thought it worth investigating. The Southern Pacific Railroad would soon be completed from El Paso to the Coast. It would make copper ore worth much more than when it had been hauled six or eight hundred miles across the desert by mule or ox team. Zeckendorff played his idea hard, and through it at one time or another owned many of the great Arizona copper mines. True to his race, he could not resist a profit. He sold all the mines too soon and made only a few hundred thousand dollars for himself and his nephew Albert Steinfeld when he should have made many millions.

Just after Warren had found the copper in the Mule Mountains a Pennsylvania engineer named Ed Reilly came through Tucson. Like everyone else he went to Zeckendorff's store to get supplies and information. Zeckendorff liked him and offered to pay the expenses and go halves if Reilly would visit the new camp and buy some of the good copper ground. Reilly took him up. A short examination made Reilly enthusiastic about the prospects. The money required proved to be more than Zeckendorff wanted to spend. Reilly had no money, but he induced two friends from San Francisco named Martin and Ballard to buy the Copper Queen claim for $20,000 and to join Zeckendorff and himself in trying to make a copper mine out of it. George Bisbee of San Francisco put up much of the money, and in return they named the settlement for him.

Reilly and his friends started to work on the outcrop near the gulch. A few feet below the surface their shaft ran into a beautiful body of 20 per cent copper carbonate. They built a little water-jacketed furnace, cut all the oak and pine trees off the Mule Mountains to make charcoal to run it, and started to make copper bullion. It was a hard game, as most of the time there wasn't enough water in the gulch to keep the jackets cool. Success looked pretty doubtful until they got two good Tombstone men who knew the desert country to run the outfit for them. Lewis Williams kept the smelter going while his brother Ben Williams handled the mine. Between them they made a pretty fair profit as long as the ore lasted.

Reilly had known James Douglas back in Pennsylvania. When he heard that Douglas was in Arizona, he invited him to Bisbee for a visit. Douglas thought he might as well have a look at the new camp, since the Verde did not need any more of his attention. The high-grade ore in the Copper Queen appealed to his imagination at once. There ought to be a lot more of it somewhere in the Bisbee limestone. Prospectors had also found a little ore on the adjoining Atlanta claim, higher up on the hill. After careful examination Dr. Douglas bought the Atlanta for the Phelps Dodge Company, paying $40,000 in cash. With the little streaks of ore to start on he began a three years' campaign of development.

Stamped by the deeply ingrained culture of many generations, Dr. Douglas must have seemed ridiculously out of place in the raw border community. The nearby silver camp of Tombstone was then at the height of its desperado days. And the scum from Tombstone took the trail over Mule Pass into Bisbee. With the Mexican border only nine miles away the hardest characters on both sides of the line made the new camp their haven. The early citizens of Bisbee were a tough lot, and they looked it.

The town was just as tough-looking as the people in it. There was only room for one narrow street along the bottom of each of the two cañons that came together just above the little smelter. Tombstone Cañon was lined with the better-class saloons and gambling houses, with a few general stores scattered among them. Brewery Gulch held the cheaper saloons and dance halls, ending in the most squalid red-light district on the border. A little farther up Brewery Gulch a Mexican farmer kept a few disconsolate pigs. This gave the group of unpainted shacks that held the ladies of badly soiled virtue who had become too battle-scarred for the more civilized Arizona towns the appropriate name of "The Hog Ranch."

Both of the Bisbee streets were two feet deep in black mud in the rainy season, and just as deep in gray dust the rest of the year. All day and all night they were filled with half-drunken miners, mule teams hauling charcoal to the smelter, and Mexicans with strings of burros delivering wood and water and supplies to the houses on the side hills. Every green thing was soon cut off or killed by sulphur smoke. Ugliness and desolation were on all sides. It was really a blessing when the choking fumes from the smelter hung low on a cold winter morning so that you couldn't see ten feet ahead of you.

Saloons and gambling houses and stores took up all the room in the narrow gulches. The little one-story shacks that housed the Bisbee miners and businessmen were perched in niches cut in the bare mountainsides. The dwellings were reached only by burro trails. There was soon a rigid social segregation. The superintendents, professional men and higher-class gamblers lived on Quality Hill, west of Tombstone Cañon. White miners and those who worked in the small stores and saloons climbed the long steps to Tank Hill, between Tombstone Cañon and Brewery Gulch. Across the gulch were the Mexicans, living like troglodytes in half-underground adobe shacks that were hardly more than pock-marks on the red slope of Chihuahua Hill. The mountain sides were so steep and the houses so close together that it wasn't safe for anyone to spit tobacco juice from his rear window for fear it would go down his neighbor's chimney.

To add to the excitement afforded by the constant din of the piano-players in the gambling houses and dance halls there were frequent shooting affrays outside the saloons. And every now and then a cloudburst sent torrents of water and mud four feet deep down the two streets. The occasional floods had their good points, for they were the only street-cleaning job that was ever done. Between storms the smell of the sewage and refuse in the gulches got pretty high.

It was still more exciting when every few months a rock rolled down from a thousand feet up the mountainside and cut a clean swathe through the wooden shacks.

One night Bill Thompson had come off shift at the Czar, and after a few drinks at Lem Shattuck's Capitol bar climbed

Rich ore in the Glory Hole, a few feet from Bisbee's Main Street, was mined out eighty years or more ago. Bisbee's first smelter, shown below in 1882, spewed out fumes that killed almost all vegetation on the mountains. It was replaced by the Douglas smelter. *(—Above, Copper Queen Branch, Phelps Dodge Corp.; below, Arizona Historical Society Library)*

up to his room below the Higgins Tunnel and started to go to bed. As he undressed, he got to thinking that he should have had one more drink. The idea of it tasted so good that he put his clothes on again and climbed back down to the saloon. While he was there a six-ton boulder of limestone came tearing down Queen Hill, went through Bill's roof, and landed right in his bed. Bill came home an hour later, took one look at the splintered legs of the bed projecting out from under the rock, and went back to Brewery Gulch to make a night of it. The next Sunday the minister of the Presbyterian church that Dr. Douglas built preached a sermon on the goodness of the Lord in saving Bill so miraculously. Most people thought he ought to have given the credit to Lem's good whisky, and not to the Lord.

Into this toughest of unkempt border camps came Dr. Douglas, spare and neat, with a well-trimmed little beard below his aristocratic nose and high forehead. He was quiet and refined in a community of cursing, loud-voiced frontiersmen. They ought to have run so rank a tenderfoot out of town. But instead they liked him right from the start. He wasn't afraid of all their roughness and appreciated them in spite of it. His fairness, kindliness and keen sense of humor could see the good points under the toughest exterior. He went quietly about his business, making friends wherever he went. It was not many months before Dr. Douglas was the most respected man in Bisbee. He kept his refinement and they kept their frontier rawness, but they worked perfectly together in spite of the incongruity.

The hard luck that had followed Douglas' business ventures in the East still tried to cling to him. He ran crosscuts and drifts all over the Atlanta for three years without finding anything except streaks of ore. At the end of this time the Copper Queen ore body also had played out on the three hundred level. Ben Williams hadn't a pound of ore left to send to the smelter. Douglas and Williams were both just about ready to quit when crosscuts from the two mines broke into a new ore body on almost the same day.

This new ore was on the sideline between the two properties. The ore was exceedingly irregular, first following a steep crack or fracture and then the flatter bedding of the limestone, like a flight of tilted steps. If they started to try to decide the apex rights in court the lawyers would get all the profits. Douglas and Williams were both wise enough to appreciate the danger. Instead of fighting, they combined the two properties as the Copper Queen Consolidated Mining Company, and made Dr. Douglas its president.

Between the Atlanta ore body and a new one in the Czar claim, the Copper Queen did pretty well for a couple of years. But the ore bodies both played out a few hundred feet below the surface. Douglas and Williams gave up hope again. They issued orders to stop all development work, mine out the few tons of ore remaining, and abandon the mines. It looked as though James Douglas would be hunting a job again at fifty years of age, with a very small western success added to his string of eastern failures.

At the last minute Fate relented and put Wes Howell on the job. Wes was mine foreman at the Czar shaft. He had a lot of respect for Ben Williams and Dr. Douglas. But twenty years underground all over the West had given him still more respect for his own nose for ore. Although he couldn't draw any pretty maps to show why, he was sure there ought to be another ore body under Queen Hill. Douglas and Williams had been equally sure that the ore would all be under the irony outcrops near the gulches. If it wasn't there, there could not be any more ore anywhere. They wouldn't let Wes waste money driving drifts into the barren mountain. Still Wes knew he was right. He figured he could charge the expense against one of the stopes, where no one would notice it. So he started a crosscut on the four hundred level straight under the hill. Ben Williams found out about it after a while and told Wes to stop the crosscut. Wes did, but a few days later started it again, orders or no orders. For months it stayed in hard, barren limestone. He began to be afraid he was wrong after all. But on the day when Douglas and Williams gave the order to abandon the mine, Wes Howell's crosscut broke into high-grade ore.

So the Copper Queen was not shut down after all. Soon the smelter was turning out copper faster than ever. The new Southwest ore body grew bigger year by year. Once in a while the ore pinched, to be sure. Then it widened out again, greater than it had been before. The ore bodies in Queen Hill proved to be the mainstay of the Copper Queen for forty-five years. They alone paid more than $20,000,000 in dividends. They — and Wes Howell's refusal to obey orders — made Dr. Douglas the leader of American mining engineers, and the Phelps Dodge Company a tower of financial strength.

With this wonderful high-grade ore to draw on, Bisbee rapidly grew to be the greatest copper district in the Southwest. Dr. Douglas had found himself at last. He gradually built up the Copper Queen into a model mining operation. The ore, of course, became lower grade year by year. This always happens, because it is human nature to mine the richest spots first. Dr. Douglas made up for the falling grade by increasing the size of his furnaces, so that he produced more copper instead of less and kept his cost just as low. When the mines became deeper, and the carbonate ore turned to sulphide, he introduced the Bessemer converting process that had just been adapted to copper ores in Butte. This increased his profit still more. After fifteen years, when the Copper Queen had filled the Phelps Dodge treasury with many million dollars, he built his own railroad two hundred and fifty miles to El Paso, and bought the "two streaks of rust and a right of way" that ran from there to the coal fields of northern New Mexico, that furnished coke for his smelter. He was on his toes all the time to improve his engineering practice and cost.

Before it approached the inevitable end of all mines the Copper Queen paid the Phelps Dodge Company a hundred million dollars in profits and gave the world two billion pounds of copper. For decades it was one of the two or three richest copper mines in the world.

The miners who worked for the Copper Queen shared in the prosperity. Dr. Douglas always led the way in paying high wages and in friendly relations with his employees. He made Bisbee more fit to live in, too. There wasn't room for a lot of civic planning, down in the narrow gulches. However, he built a good library and a hospital and a clean hotel, and finally helped to pave the streets so that people no longer had to wade ankle deep in dust or mud every time they bought a loaf of bread or took a drink. After a few years he moved the smelter to the new town of Douglas, in the Sulphur Springs Valley. With the pall of sulphur smoke gone, trees and flowers grew again in the cañons and on the mountainsides. Gambling was made illegal, and the gunmen moved away or got respectable. Bisbee lost some of its picturesqueness, but it was no longer the first-class cross section of hell it had started out to be.

His company and his fellow engineers recognized Dr. Douglas' accomplishments. The success of the Copper Queen before long made him the wealthy president of the Phelps Dodge Company. A little later he was elected president of the American Institute of Mining Engineers. All the honors the mining profession had to give were heaped upon him. For the last thirty years of his life he was almost revered by all who came in contact with him. Whether he had made the Copper Queen or the Copper Queen had made him, the combination of man and mine was one of the most fortunate in all the history of copper.

In spite of the technical skill and foresight that he had acquired since the days of the Quebec Presbyterian Church, Dr. Douglas had not learned to see into the ground. He gradually bought all the claims he thought he might need. But in Bisbee the rocks on the surface gave few clues as to where the ore would be. By the time he had extended the Copper Queen property southeast as far as the jagged brown cone of Sacramento Hill, he was satisfied. This old volcanic neck and a limestone ridge west of it seemed a logical place for the Bisbee ore bodies to end. Prospectors tried to sell Dr.

Douglas ground still farther south. He said the Copper Queen had all the ground it needed. If anyone could find ore under those barren lime hills, he was welcome to it.

This proved to be Dr. Douglas' second big mistake, even more expensive than turning down the United Verde.

A wild-eyed, drunken Irishman called Jim Daley had located the ridge west of Sacramento Hill early in the 'eighties. He called the claim on top of the hill the "Irish Mag," in honor of one of the famous old girls of The Hog Ranch. Daley did a little shallow prospecting up on the Mag Hill without finding any ore. Most of the time he spent fighting the Copper Queen about a right of way. He wanted to cross their ground one way, and they said he had to go another way. To maintain his rights, one day he beat up a Mexican who tried to keep him off the forbidden trail. Bill Lowther, the local constable, tried to arrest Daley. Daley shot him dead and fled across the border into Mexico. There he died himself, after a particularly satisfying drunk.

The estate Daley left did not look very valuable at that time. However, a whole flock of heirs appeared to claim it. A Mexican woman who had been living with him in Bisbee said she was his widow, and sold her rights for $1,800 to Jim Cunningham. Cunningham was prospecting on a grubstake for Martin Costello, the leading Tombstone saloonkeeper. Another Mrs. Daley and her son came down from Leadville, Colorado, when they heard that Jim was safely dead. Then Andy Mehan, who ran a little saloon in Brewery Gulch, produced a bill of sale that he said Daley had given him in Mexico just before his last drunk. Two Tombstone storekeepers named Cohn foreclosed a mortgage on Mehan, and assumed his rights to the Daley property. The whole crowd of them fought it out in the courts, to the joy of the lawyers. It took ten years to wind the case up. Finally Martin Costello won out, and got title to the Irish Mag and four other claims south of it.

All the time the suits were being tried, the Copper Queen kept finding new ore bodies farther southeast, toward the Irish Mag sideline. By 1898 the ore in the Silver Spray – the last

This view of Bisbee in the 1890s looks up Tombstone Canyon. Below, the hoist of the Copper Queen's Spray Mine, about 1902, used flat ropes and reels. *(—Both, Copper Queen Branch, Phelps Dodge Corp.)*

Copper Queen claim — was only a couple of hundred feet from the line. Dr. Douglas was weakening a little in his opinion. Still the chance looked too slim to justify the big price that Martin Costello now asked.

Bisbee had by this time become one of the famous copper districts. Mining men from other camps often dropped in to see the rich limestone ore bodies that Dr. Douglas had developed. Among them, early in 1898, were two Calumet and Hecla foremen or "mine captains" off on a vacation. "Cap'n Jim" Hoatson was the best miner in the Copper Country. His younger brother "Cap'n Tom" was nearly as keen. In the Bisbee saloons they found a lot of old "Cousin Jack" friends who had worked for them in Michigan. From them they learned that the drifts running toward the Irish Mag sideline on the six hundred level of the Spray were in high-grade ore.

This sounded interesting. Cap'n Jim found that the Irish Mag group was for sale. The price of $550,000 cash that Martin Costello asked seemed ridiculous for a property without a pound of ore in sight. Still it would not take much 8 per cent ore like that in the Spray to be worth it. Cap'n Jim determined that he would not leave without looking the ground over carefully.

The two mine captains spent the next few days asking all their old miner friends just how the Bisbee ore occurred and climbing over the sharp gray limestone beds on Mag Hill to see if they could find any clue to possible ore bodies. It was a hard problem for two Cornish miners,* who had worked their way up to be Calumet and Hecla foremen without the help of any school learning. There was mighty little sign of ore on Mag Hill. All they could see was a lot of seams of iron oxide, with a little copper stain, cutting across the limestone beds. But then there wasn't much more sign of ore over the Spray bodies. Practical miners like Cap'n Jim and Cap'n Tom would have been horrified if anyone had accused them of being geologists. Yet this was a very complex geological prob-

*Cap'n Tom's son called my attention to the fact that his father and uncle were Scottish miners, not Cornish.

lem they were up against, and they had to use geological reasoning to solve it.

The rocks on Mag Hill were steeper and rougher than anything Cap'n Jim was used to up in the Lake Superior country, and thirty-five years underground had taken away some of the resiliency from his short legs. One day he sat down to rest on a chunk of limestone beside the brilliant red blossoms of an ocatilla. For half an hour he stroked his pointed white beard and tried to figure out what it all meant. The desert sun was hot, even in February. He wasn't used to working his head so hard. Gradually the outline of the red hills across the cañon grew hazy. Before he realized it, he had drifted into a doze. As he slept a vision came to him, clear as the mine maps he could read so easily. He saw a great bed of ore beneath his feet. Nine hundred feet underground, it followed the layers of limestone from the Spray clear across the Irish Mag. Hundreds of feet wide and eighty feet thick, it was greater and richer than anything he had ever imagined. And the surface outcrops showed just where the ore would be. The intersecting iron seams had leaked up along cracks in the limestone directly above the great ore body. They were worthless themselves, but they outlined the ore perfectly.

This vision of Cap'n Jim's was the beginning of the Calumet and Arizona. It was so clear that even when he was fully awake, he did not question it for a minute. All he had to do was to sink a nine-hundred-foot shaft, and the ore body was his.

The Hoatsons went back to Calumet eager to raise the money and start sinking. Everyone in the Copper Country loved and trusted Cap'n Jim. They would follow him, vision and all, rather than the most famous engineers. The leading banker of Calumet, Charles Briggs, contributed heavily and became president of the new Superior and Western Copper Company. All of the Calumet and Hecla bosses and their friends bought what stock they could at sixty cents a share. Before the end of the year the new company paid Martin Costello his $550,000 and started to sink the Irish Mag Shaft.

Cap'n Jim and his friends had not realized that mining down in the Arizona desert cost much more than it did in Calumet. The money they had raised for the Superior and Western was far too little. By the time the shaft was seven hundred feet deep they had only a few hundred dollars left. Cap'n Jim hated to quit before he reached the depth his vision had indicated. But there was no way out. With the last few dollars he ran a crosscut on the seven hundred fifty level over toward the Spray. It found only narrow stringers of ore. Superior and Western seemed doomed to failure. The Irish Mag was shut down, and the stock dropped to a few cents a share.

Still Cap'n Jim's friends had not lost faith in him or in the mine. One of the best of them was Thomas F. Cole. Tom Cole had fought his way up by grim determination and hard work from digging post holes in the Calumet district to the presidency of all the Steel Corporation iron mines in the Lake Superior country. When he started a thing he never quit, whether he was right or not. He had joined Cap'n Jim in the Superior and Western, and he would see him through.

Through his iron mining, Tom Cole knew many of the Steel Corporation leaders. He persuaded Henry W. Oliver of Pittsburgh that the Irish Mag was a promising venture. Oliver and his friends Chester A. Congdon and George E. Tener, both successful in the iron and steel industries, agreed to advance the necessary money to save the almost bankrupt Superior and Western Company. They reorganized it as the Calumet and Arizona Mining Company, and started development again in the Irish Mag Shaft.

On the nine hundred fifty level they drifted out to where Cap'n Jim had seen the ore in his vision. The drift cut three hundred feet of high-grade copper ore. Just as Cap'n Jim had dreamed, it extended in a great layer clear across the claim. Tom Cole and the Hoatsons and all their friends became rich men almost over night.

The Calumet and Arizona hastily built a smelter in Douglas and started to mine the Irish Mag ore late in 1902. A few months later it paid its first dividend. The ore body that re-

BEE DISTRICT ARIZONA
OLIVER MINE
IRISH MAG MINE
BISBEE
GARDNER MINE
LOWELL MINE

These panoramas of the Bisbee district and its principal mines, looking north, and of the Junction Mine plant were both taken in 1916 for a pamphlet prepared for a meeting of the American Institute of Mining Engineers.

sulted from Cap'n Jim's vision was the richest Bisbee ever saw. Before its wonderful story was done, the Irish Mag paid $15,000,000 in dividends from its fifteen acres. C. and A. stock paid the old Superior and Western investors more than a hundred for one.

The success of the Irish Mag was soon followed by ore bodies only a little less valuable in the Oliver Mine of the Calumet and Arizona, still further south. Then Lem Shattuck of the Capitol Saloon started a little development company in the hills west of the Copper Queen, and found bonanza ore. Harvest time had come for the prospectors, as they could sell the most barren looking claims for fancy prices.

Dr. Douglas realized now that ore would continue far out under the limestone ridges south of Sacramento Hill. He determined to get his share of it for the Copper Queen. Tom Cole and the other Calumet and Arizona directors were just as eager to repeat their spectacular success. Engineers for companies in other districts tried to get promising claims. A wild buying scramble put the price of absolutely undeveloped ground up to thirty or forty thousand dollars an acre. New fortunes were made every few days. Bisbee flourished in one of the greatest of all the mining booms.

Unfortunately for the companies, but luckily for the prospectors, the surface gave no hint whether a claim was a good or a worthless one. South of Sacramento Hill the ore was covered by increasingly deep layers of barren limestone. No one could guess where the ore bodies were going to be. The only way to find out was to sink a ten- or twelve-hundred-foot shaft and then to spend a few hundred thousand dollars in horizontal crosscuts from it. The claim-owners insisted on cash before the work was done. It took a two- or three-million-dollar ante to get into the Bisbee poker game. The outside companies found it too steep and soon withdrew. The Copper Queen and Calumet and Arizona fought it out with such varying success that the Bisbee property map soon looked like a patchwork quilt.

The Lowell Mine was the first prize they went after. This struggling development company owned four or five claims east of the Oliver Mine. Senator Clark had taken an option on the Lowell soon after the Calumet and Arizona found its first ore. He sunk an eight-hundred-foot shaft without finding anything save lean iron sulphide. Then he threw up the option. This gave Dr. Douglas a chance to get at least partly even for his wrong guess at the United Verde. The Lowell owners wanted half a million dollars in cash. Tom Cole and his associates tried to beat them down. While they were haggling, Dr. Douglas paid the half million. Deeper development soon found ten or twelve million dollars' worth of ore. For one of the few times in his life, Senator Clark had been too conservative.

The Del Norte claim lay east of the Lowell. The Calumet and Arizona group weren't going to lose another good mine by being too slow. They cheerfully paid nearly $500,000 for the sixteen acres. Dr. Douglas paid nearly as much for the Hardscrabble, still farther east. Martin Costello and Billy Brophy and the others who had backed the prospectors in the early days raked in money so fast they did not know what to do with it.

Tom Cole looked on it as a personal affront that the Copper Queen had bought ground he wanted. He would not let that happen again. Taking charge of the campaign himself, he soon knew more about the hundreds of outlying claims than anyone else in Bisbee. (After thirty years he could still show the engineers the corners of every claim and tell who owned it and how much it cost.) He took options for prices that everyone else thought crazy. Backed by the prestige of the Calumet and Arizona, Cole and the other directors formed six development companies with millions of dollars paid in capital. These companies divided among them the thousands of acres of possible ore-bearing ground that Cole and his associates had optioned. The Calumet and Arizona crowd at last had the Copper Queen hemmed in on all sides where extensions of the ore were possible.

Dr. Douglas might still have taken away most of the ore from the intruders. The limestone beds in which the ore occurred, as well as the ore itself, came to the surface in Copper Queen ground. A Heinze would have claimed an apex on all the Bisbee ore – and might possibly have got away with it. Dr. Douglas very wisely recognized that ore occurrence in Bisbee was so complicated that only a ridiculous twisting of words could force it under the apex law. If he started suit, the litigation would last for many years. The resulting uncertainty, expense and hard feeling might wreck either or both companies. Therefore he signed an agreement with the Calumet and Arizona making all sidelines vertical. Owing to this agreement, the Bisbee miners lived together in friendly rivalry. In thirty years of mining, in spite of interlacing property lines, there was not a single lawsuit between the companies. All disagreements were settled by conferences among the engineers. Representatives of either company were always welcome to visit the other. Nothing like this had ever been known. In all other districts visitors were as popular as rattlesnakes.

The policy of friendly coöperation gradually spread through Arizona, then through all the West, and finally through the whole mining world. The interchange of information that resulted has been largely responsible for the rapid growth of the arts of mining and metallurgy. The wisdom of Dr. Douglas in refraining from apex suits in Bisbee may have cost the Copper Queen some ore. It was worth many times the value of this ore in the technical progress that it brought about. The good feeling and coöperation that exist among engineers are perhaps the finest monument Dr. Douglas has left.

After the ground had been bought and protected by vertical sideline agreements, the new children of the Calumet and Arizona were faced with enormously expensive development. The farther from Sacramento Hill the ground lay, the more it cost to find the ore. As shafts grew deeper, staggering flows of water were encountered. The porous limestone was like a great lake. Pumps that cost hundreds of thousands of dollars lifted twenty or thirty thousand tons of water a day from twelve-hundred-foot depth. Money poured in as fast as the water poured out. It looked for a time as though all the Calumet and Arizona profits would go back in the ground again.

In the end Tom Cole's optimism was more than justified. One by one the new mines found great ore bodies. It took $4,000,000 to do it at the Briggs Mine, but at last there, too, a wonderful body of high-grade copper ore rewarded the expenditure. The most daring campaign of underground development ever undertaken was a complete success. Before the ore bodies in the Irish Mag and the Oliver were exhausted, the Calumet and Arizona absorbed the new development companies and was assured a life of fifty years.

Captain Jim Hoatson's vision had already yielded more than a billion pounds of copper and $62,000,000 in dividends. It built Bisbee up into a prosperous city of eighteen thousand inhabitants. And the mines Cap'n Jim started never had as much rich ore developed as they did after thirty years of intensive operation.

Tom Cole did such a good job in assembling the outlying claims that at last the great Copper Queen had no place to turn for new ore. It must either shut down or buy the Calumet and Arizona. The price was a high one. For their Bisbee and Ajo mines the Calumet and Arizona directors insisted on a valuation equal to that of all the far-flung enterprises of the Phelps Dodge Corporation. At last the Phelps Dodge agreed to pay the price.

After a fine life of daring success Calumet and Arizona has ceased to exist. But the mines that it developed insure long prosperity to Bisbee and the company that is a worthy offspring of the chance union of Dr. Douglas with the Copper Queen Mine.

AUTHOR'S NOTE: Both Dr. James Douglas' son, James S. Douglas of United Verde Extension fame, and his grandson, Lewis W. Douglas, former ambassador to Great Britain, have maintained that the story of Dr. Douglas' "turning down" the United Verde was false. These two good friends prefaced every meeting I had with them over many years with an insistence on the incorrectness of this flaw on the escutcheon of one who was probably the best mining engineer of his generation. Lewis Douglas went to the trouble of searching through fifty-year-old files and found Dr. Douglas' original report on the United Verde. It said the prospect was a promising one, but the remoteness of Jerome would make it impossible to operate successfully until there was a railroad in northern Arizona. When the tracks were being laid from Ash Fork to Phoenix, Dr. Douglas tried to acquire the United Verde, but Senator Clark was a little ahead of him. I am glad to apologize to the Douglas family for my part in spreading the widely accepted story that Dr. Douglas turned down the United Verde.

Bisbee has lived up to the hope that it would have a long and successful life. As outlined in the chapter in Part II on the great low-grade copper mines, the vision and daring of President Louis Cates and Manager Harry Lavender resulted in a great low-grade open-pit copper mine under and south of the brown pits of Sacramento Hill. Sad to say, this low-grade deposit is comparatively small and is approaching an end. The rich limestone-replacement ore bodies also cannot long stand greater depth and increasing costs. Unless there is a fantastically high copper price, after 90 prosperous years Bisbee must soon join the dismal list of "ghost towns."

6

Bill Greene of Cananea

It isn't often that a rag-tail gambler gets a chance to play for hundred-million-dollar stakes. But now and then Fate gets tired of building civilizations and toppling them over again, and gets up a vaudeville show of all the most unlikely events she can think of. A lot of relaxation was in the cards after all the hard work of starting up the Industrial Age in the 'eighties and 'nineties, so Fate turned Bill Greene loose to entertain her.

Bill had spent forty-seven years being just as worthless and ornery as the Lord would let him. The little Upstate New York town of Hornellsville was too quiet for a boy who wanted to get some adventure out of life. So Bill drifted West and wandered around the frontier, driving teams for the Government in Kansas, fighting Indians and prospecting in the Bradshaws in northern Arizona, and punching cows for fifteen dollars a month whenever the cards had turned against him. He even got down to chopping wood and packing it into Prescott on a burro to sell for a few dollars a cord, and only the Mexicans are supposed to do that. Money just wouldn't stick to his fingers as long as a faro game was within reach.

Along about 1890 Bill was close to forty years old, and it wasn't so much fun sleeping out on the desert with nothing but a blanket, listening to the coyotes yowling and wondering where his next dish of beans was coming from. He figured it was about time for him to settle down. Tombstone was going good then, and any miner could get five dollars a day. It was tough on Bill to keep away from the Bird Cage and the other dance halls and gambling houses, but he stuck with it until he saved money enough to buy a little adobe ranch house down on the San Pedro near Hereford. There was lots of good open range in southern Arizona in the 'nineties, as barbwire fences were still cause for shooting. All a man needed was a few cows to start with and nature would do the rest. Bill was handy with the running iron after all his years punching cattle for other fellows, and he soon got together a nice little herd. They hadn't cost him much, but it wasn't healthy to ask a new cattle man in that part of the country just how he got his start. So his neighbors left Bill alone, and he got married and started to raise a family and watch his calves grow up to the time when they could keep him in gambling money.

As long as there was plenty of elbow room between Bill and his neighbors, he got on with them all right. But Jim Burnett was a little too close by. He was Justice of the Peace at Hereford, and he had a place down on the San Pedro just below Bill's. Lots of times there wasn't enough water in the river for them both to irrigate their little alfalfa patches with, and they used to cuss each other pretty lively when they both wanted it at once. It really wasn't worth fighting about, but neither of them was any good at forgetting a grudge. Anyone could see that sometime when they were both over in Tombstone having a few drinks there was going to be trouble. Trouble in those days meant six-shooters and a new grave on Boot Hill.

The main difference between Bill Greene and the other prospectors and would-be cow men down on the border was that he could not see things the way they were. When he sat on his heels and eased his big shoulders down against a cottonwood tree and pulled his long moustache and looked over at the Huachucas, he didn't see just the brown grass-covered desert rising up through the heat waves mile after mile to the blue wall of the mountain. What he saw was a big dam there in Miller Cañon, with ditches leading down to a lot of alfalfa fields and orchards with clean white adobe houses sitting in

"Colonel" William C. Greene, as he looked at the peak of his career, made Cananea a great mine and started several ambitious ventures that others later made successful. *(—Arizona Historical Society Library)*

the midst of the oleanders. His neighbors all thought he was crazy, but he paid for his share of the drinks and always had some beans ready for a friend who needed them, so they bore with him.

This inability to leave the desert alone nearly finished the story for Bill on the end of a rope before his real play began. He wasn't satisfied with his little patch of alfalfa down in the San Pedro bottom and decided to irrigate some bigger fields to help his cows through the dry season. So he built a dam of brush and mud to divert the trickle of water into his ditch. This was all right for a few months. But when the rainy season came along there was a cloudburst up in the mountains and a wall of water four feet high rolled down the San Pedro. That was the end of Bill's dam. That afternoon Bill's little girl was playing in the sandbank a little farther downstream. It was peaceful and sunny down there, and she didn't notice the black thunderheads around San Jose Mountain. When the flood took out the dam it rushed right down on her and drowned her before she had any idea what was coming.

Bill thought an awful lot of his little girl, and when a cow hand brought her dripping body up from the river bed he just went crazy. The first thing he thought of was Jim Burnett. The Judge had said he wasn't going to stand for Bill keeping the water from coming down to his place, and Bill jumped at the idea that Burnett had blasted the dam and drowned his little girl. He was too upset to think that a flood like that would carry away the dam without any help. He didn't say a word to anyone but just climbed on his horse and rode the eight miles across the bare hills to Tombstone as though the devil were after him. When he got there Judge Burnett was just walking out of the O K Corral. Bill jumped off his horse and shouted, "Vengeance is mine: I will repay, saith the Lord." Before the Judge could catch his breath to ask what was the matter Bill pulled his gun and shot him through the heart.

It looked like a cold-blooded murder, and a lot of the boys wanted to string Bill up. If Sheriff Scott White had not been a good friend of his, Bill wouldn't have had a chance to try out any of his other big ideas. The Sheriff wasn't afraid of any mob. He held them off with his six-shooters until he got Bill safe behind the bars in the jail. And then at the trial White and some other friends of Bill's swore that Burnett had threatened to shoot him on sight, and the jury called it self-defense and let him off. It was a close call for Bill, and he never forgot Scott White or the other friends who saved him.

After that Bill had enough of improving on nature for a while, and he stayed pretty close to his ranch except when he had sold some steers and could go up to Tombstone or Bisbee and do a little gambling. But along about 1898 his feet began to itch for a game with bigger stakes than the price of a few calves. The papers that filtered out to Arizona were full of big market plays that Henry H. Rogers and Tom Lawson and Heinze were making in Butte copper stocks in Boston and New York. Bill figured that he knew more about gambling than they did, and all he needed was the price of an ante and he could run them right out of their own game. So he set out to find himself a copper prospect that would beat the Butte and Boston and all the other big mines that were catching the millions of the eastern suckers.

This looked like a tough contract for a forty-seven-year-old cow man who had never been more than about two jumps ahead of the poorhouse. But luck played with him. The showing he was looking for was waiting for him only a few miles from home, and he found it at just the right time. The cards had taken a long time to come his way, but when they did he got a hand that no one could beat.

The Mexican border was only about four miles south of Bill's ranch house. There wasn't any line fence in those days, and his cows could wander south just as well as any other way. There the San Pedro and the Sonora River both headed in a broad mesa of grassy ridges dotted with live oaks. It was the sweetest cattle country in all the West, and Bill's steers knew it. Every few months he would have to ride across the border to round them up and drive them back home. One day, just about the time he got it in his head he wanted a copper prospect, the steers had wandered farther south than usual. He followed them for nearly forty miles clear to the place where the mesa ended in the steep oak- and pine-covered slopes of Cananea Mountain.

Bill forgot all about the steers then. For the rocks and soil that showed through the brush on the north side of the mountain were bright red and brown. There were miles and miles of the brilliant stain. It covered the whole slope from Puertocitas Cañon clear to the east shoulder of the mountain. Bill had seen staining like this around the big copper camp of Bisbee, just north of the border. He knew it meant iron sulphide and that where there was a lot of iron in that part of the country there was copper too. This stain covered ten times as big an area as that around Bisbee, and it was even redder. If the copper was on the same scale as the iron, he had the biggest mine in the world.

It didn't take him long to find out that Cananea Mountain had plenty of copper in it. There was a little Mexican settlement called Ronquillo at the foot of the mountain, just east of Puertocitas Cañon. Bill tied up his horse at the best house in town, which was a clean white adobe surrounded by bright-colored flowers. A pleasant looking middle-aged Mexican woman came to the door when he knocked and asked him to come in out of the sun. Bill talked good border Spanish, and he always had a way of making the Mexicans like him — maybe because he was friendly and square with them instead of treating them like dirt the way a lot of Americans did. Before he had been in Ronquillo an hour he knew all about the lady whose house he had gone to and everything that had happened there since the settlement was started. Bill was a big, good-looking gringo in spite of his forty-seven years and the long ride he had had, and she was glad to get a chance to talk to someone beside the handful of peons who made up the village.

Bill found that she was the widow of a Mexican general named Ygnacio Pesquiera. He had come to northern Sonora in 1865 with five hundred soldiers to help clean out the

Apaches, who were raising hell as usual. While he was camped near Cananea Mountain some prospectors brought him chunks of rich copper ore that they had found in "antigua" prospect holes left by the Spaniards of two hundred years ago. Pesquiera had done a little mining in other parts of Mexico, and this ore looked good to him. So he made his headquarters at Ronquillo and built a little adobe furnace to melt down the ore in. Whenever his army was not fighting Apaches he had the soldiers leasing on the claims he located back on the mountain. They sorted out the high-grade copper ore and packed it down to the smelter on burros. There General Pesquiera melted it with charcoal to a high-grade copper matte, which he packed by mule train three hundred miles to Guaymas, on the Gulf of California, and from there shipped it by sailing vessels around Cape Horn to Swansea, Wales. It was a complicated way to handle it, but the ore was so high grade that it made the Pesquiera family the richest one in all that part of Mexico. Mining started them, and cattle ranching did the rest.

General Pesquiera ran his copper smelter for about fifteen years. Then he died. Ever since the mines had been shut down, except for the "gambosinos" who now and then stole a little high-grade ore. Back in the 'eighties an American company had tried to smelt some malachite ore up at Puerocitos. The ore was too low-grade, and the company soon failed. The widow Pesquiera was getting pretty tired of paying taxes. If her distinguished visitor "Meester Greene" could be prevailed on to enter the copper-mining business, she would be delighted to let him have her "Cobre Grande" for a mere pittance.

This was sweet music to Bill's ears. He hardly needed to go up to the old workings to see the veins of rich copper glance ore that General Pesquiera had left. The thousands of acres of brilliant red stain were enough. As Bill sat in the cool patio and sipped his mescal, his imagination sank shafts into gigantic bodies of high-grade ore. After another drink a great smelter was belching forth columns of billowy smoke, while a continuous stream of white, molten copper poured into the long line of molds. No rich easterner could possibly resist the glowing story he would tell of the hidden wealth of Cananea, waiting only for the magic touch of Bill's genius and a million or so of the other fellow's money to blossom into dividends that would beggar Croesus himself. A third drink and his mind was made up. Before he stumbled off to sleep in Señora Pesquiera's best bed, Bill had taken an option on her mining property for $47,000, with a cash payment of promises.

The millions were almost in reach of his fingers now. Only he had to have a few hundred dollars cash before he could make the big play. That meant partners, and he knew just the right ones. Jim Kirk came first of all. He was the best miner in Tombstone, and when he put on his long Prince Albert coat and two-foot black sombrero and smoothed down his great drooping moustaches, anybody in the Southwest would come a-running to get into the same company with him. Jim was the squarest man in the country, too, which was good for Bill but might be inconvenient if he did too much talking to stockholders. Best of all, Jim could get a job any day as foreman in Tombstone or Bisbee, and the hundred and fifty a month would keep the partnership in beans, anyhow.

Ed Massey was the one Bill chose for the third partner. Ed had done pretty well as a miner in Tombstone and Bisbee, but his long suit was talk. When he opened up with his promises, he could charm the orneriest Mexican miner into pounding the drill for another week before payday came. Paydays were likely to be scarce until Bill really got his big mine sold, so Ed was just the man to run the prospecting crew. As long as the promises and the beans held out, the dozen Mexicans they set to work sinking on the best showings would never quit.

With the partnership all formed and development started, the next thing was to get a report they could sell the mine on. This was Bill's job, and he sure could fill it. After thirty years in and out of Tombstone and Bisbee and the Bradshaws and a half dozen other mining camps he knew all the long words that any of the experts could sling. With his imagination he could put them together so they would charm the dollars right out of the pockets of Russell Sage. So Bill set to work. All day he scratched his head and pulled his moustache and wrote at top speed. When evening came and his partners had dropped in to hear the results of the labor, he threw down his pencil and said, "There! Thank God, she's finished, and she's a daisy." He just had to read it over once more before they could see it. So he read a page and hitched around in his chair; and read another page and bit off another chew of tobacco; and read some more and ruffled up his gray-brown hair, getting more and more excited until at last he tore the report violently in two and threw it on the floor. "What's the matter, Bill? Are you crazy?" Jim asked. "No, by God. That mine's too good to sell. I'm going to keep her!"

Next morning Bill had sobered down and was ready to sell the mine again. He formed the Cobre Grande Copper Company, and wrote another report and went East to sell stock. Jim Kirk was hard at work in Tombstone earning the expense money, and Ed Massey kept the Mexicans down at Cananea hoping that next payday they would get something better than credit for beans at the store. The prospect shafts were turning out fine, too. They were finding 5 or 10 per cent copper ore wherever they went in the Oversight and the Veta Grande. Bill did not have any trouble talking about millions of tons of ore now. He was so big and handsome with his quiet voice and gray hair and brown moustache that people just naturally had to believe him. Early in 1899, only a few months after he had taken the option from old lady Pesquiera, Bill turned over the company to J. H. Costello of Philadelphia and George Mitchell of Jerome, for the promise of two hundred and fifty thousand in cash and a big stock interest.

This would have been a pretty good play if it had gone through. But Bill always guessed wrong when it came to picking men, and instead of getting his big stake he pretty nearly lost his shirt. Mitchell came down to take charge of the property, and started to build a two-hundred-ton smelter. But he was long on talk and short on action. There was no sign of the two hundred and fifty thousand, and Bill soon saw that he was going to be eased out without getting any cash or stock or anything else. Mitchell and his superintendent, Con O'Keefe, had possession and they let Bill know that he was just an outsider who could go rustle for a living for all they cared.

Bill came awful close to being the sucker this time. But he had an ace in the hole all ready for any such emergency. He would stand back of his word to a friend to the limit, but a business deal was a gamble. Every man had to protect himself when it came to making contracts. So just to make sure no unpleasantness came up, he had omitted to record the transfer of his option on the Cananea claims to the Cobre Grande Company in the Mexican district office in Arizpe. Bill knew that this made the transfer no good. By this time, "Meester Greene" was the best friend of all the Mexican officials in that part of Sonora. The local court was delighted to oblige him by throwing Mitchell and all his crowd out and handing the property back to Greene unencumbered. Instead of getting his $250,000 he had the mine with a lot of new ore developed and with a two-hundred-ton smelter.

This was a good start. But to make any real money he would have to build a mill and a much bigger smelter and a railroad, to say nothing of shops and a town and any number of other expensive incidentals. It was a million-dollar job at least.

Jim Kirk came down to run the mine this time, as it had got too big for Massey. Bill formed the Cananea Consolidated Copper Company and went back to New York to raise his million. The little smelter was turning out copper fast now, and he could get a few thousand dollars credit at the banks along the border. He would show the dudes in New York what spending was. He hired a fine suite at the Waldorf and bought a full outfit of the best clothes money could buy. Flowing black frock coats and big black sombrero, Arizona gambler style, made all the New Yorkers ask who was the distinguished visitor. Five-dollar tips made the bellboys give glowing answers. Every afternoon when the brokers stopped at the Waldorf on their way up town, Bill would stand up at the big horseshoe bar and say the drinks were on him. They laughed at him at first but they took the drinks and pretty soon they got to know him and to like him. Within a month he was one of the big celebrities. It was Colonel Greene of Cananea now – the great new copper magnate.

Still the million kept just out of reach. Bill talked his best to Henry H. Rogers and William Rockefeller down at the Amalgamated office, but they figured his story was too good to be true, and threw him out. His broker friends liked to listen to him, but they had other business when it came to buying Cananea stock. Bill was spending a lot more than Jim Kirk could make out of the little smelter down in the mountains forty-five miles from the railroad. Every day the expenses got higher and the credit lower. At last Bill was down to the last hundred Kirk and Massey could beg or borrow from the banks and their friends down on the border. Anyone but a gambler would have been licked. But Bill knew his luck must be about ready to turn, as he took the last hundred into Canfield's gambling house, where all the New York swells did their playing. The cards came his way, and he pressed his luck for all it was worth. That evening he came out with $20,000 safely tucked under his belt.

Bill knew he would come out all right now. He started to spend and to talk more than ever. Soon he located a man who was just as big a gambler as he was. Tom Lawson recognized a kindred spirit and agreed to honor Bill's drafts up to a million dollars in return for short-term notes and an option on control of the six-hundred-thousand-share Greene Consolidated Copper Company at a third of the ten-dollar par value.

So at the end of 1899, Bill went back to Cananea with his pockets full of money. And he surely made it fly. Those were the best days the desert had ever seen. Bill had fifty big mule teams hauling supplies and machinery down from the railroad at Naco, on the border. He hired all the American miners and machinists and carpenters he could get at five to eight dollars a day. Men were so thick they were falling all over one another. The machinery salesmen were in heaven. Bill would buy on the spot all the hoists and compressors they offered him. The clerks and timekeepers had a hundred and fifty fine saddle-horses to ride around on, keeping track of the men who were building the new thousand-ton smelter, a six-hundred-ton concentrator for the leaner ore, and the finest American town in the Southwest up on the mesa. Money was running like water. Saloon men and gamblers and dance hall girls from hundreds of miles around flocked in to get their share of it. The main street of the old Mexican town of Ronquillo down in the gulch between the mesa and the mountain was crowded day and night with drunken miners elbowing the bums and prostitutes of a dozen nations out of the way. Bill was the only law and order in town, and as long as there weren't too many shootings he turned them loose to go as far as they liked.

Old Jim Kirk kept his head and went on digging holes in the mountain. Everywhere he dug he found rich ore. The Oversight and the Capote soon had developed four or five million tons of 8 per cent ore and any amount of leaner stuff around it. On top of that Kirk ran into the daddy of all the ore bodies down in the Veta Grande. It was twelve hundred feet long and four hundred feet wide, all 6 to 10 per cent copper glance. Nothing like that had been seen since the bonanza days in Butte. Ed Massey did so much talking about it that everybody called it the Massey ore body, but Jim Kirk was really the one who found it. Cananea had made Bill Greene's wildest dreams look like pikers.

Right at the peak Bill nearly lost out again. Tom Lawson had advanced $135,000 out of the million he had promised and had taken short-term notes for it. Now he saw a chance to freeze out Bill and to get the whole property for himself. Without any warning he quit honoring the drafts from the mine and started to foreclose on the notes. Bill was spending two dollars for every dollar's worth of copper the smelter turned out, and he owed money to all the banks and machinery

Thomas W. Lawson had this portrait made in 1905, when he was fighting Bill Greene in the New York stock market for control of Cananea. *(—The Bancroft Library, University of California, Berkeley)*

men along the border. He had to act quickly or he was gone. So he hurried to New York and opened up a big office and started to sell stock. John W. Gates and his associates, Hawley and Huntington, came to the rescue just in time. Their money paid off the notes, and Bill had a good excuse not to turn over any of the stock he had promised to Tom Lawson. Instead Gates and Hawley helped him peddle it out to the public at $10.00 a share instead of the $3.33 Lawson was going to pay.

All the time the orgy of spending went on. Jim Kirk kept sending high-grade ore to the smelter, and as fast as the copper was sold, Bill paid out the returns in dividends. But for every dollar he paid he had to raise six or eight by new stock issues. By the end of 1901 he had spent six millions and had only started the improvements he planned. If the ore had been pure copper it could not have stood Bill's sort of operation. A good mine has to be able to stand a reasonable amount of champagne and bad management, but Bill was too much for any ore bodies.

The only thing he got cheaply was the forty-five-mile railroad from Naco to Cananea. He started to build in 1901, in spite of the fact that there wasn't nearly enough money in sight to finish the job. E. H. Harriman was keeping a close watch on all the railroad building in Sonora then. He figured he could pick up the Cananea line for a song when Bill went busted. On one of his trips west he had his private car set out on the siding at Naco, and asked Bill to come and see him. "You know and I know you haven't got money enough to finish this railroad," Harriman said – which was right. But Bill could still bluff, and he said, "I'll bet you a hundred thousand dollars I have," and he pulled out ten $10,000 bills. He had borrowed every cent he could just to put one over on Harriman. The game worked, and instead of taking up the bet Harriman bought the railroad for the Southern Pacific at a big price. Trains were running before the end of 1902, and Bill could turn his mule teams out to graze.

There wasn't any end to the spending, or to the ore bodies. But the ore wasn't quite as rich as it looked when Jim Kirk ran the first crosscuts through it. A lot of it really carried 8 or 10 per cent copper, but a lot more was only half that rich. The 4 per cent ore was too silicious to smelt without first concentrating it. Bill had realized this when he built the six-hundred-ton mill. Unfortunately the only mills he had ever seen were the silver mills at Tombstone. So of course the Cananea mill lost about as much copper as it saved. For once Bill saw that the job was too much for him or for his old Tombstone friends. He sent up to Globe for Dr. Ricketts to build a new mill for him, four times as big as the first one. This was the wisest thing Bill ever did. The "Doc" did not look like much in those days, with a dirty old slouch hat and soft shirt and pants that had never been pressed hanging on the corners of his bony hips as though they would fall off any minute. Even the Mexicans called him "Mal Cinto," which means "The Badly Cinched." Every now and then a new Pullman conductor on the main line would try to throw him off the train, thinking he was a hobo. In spite of his looks he was the best engineer in the business, and the Copper Queen, Amalgamated, and most of the other copper companies had to get him when there was a tough job to do. Bill hated to see the big fee go out of the family, but this mill had to work. The mill cost a couple of million dollars, as Dr. Ricketts insisted on having the best of everything. When it started up, late in 1904, it ran like a clock, just like everything else the Doctor planned. It came just in time, as the grade of the ore Jim Kirk mined dropped from 7 per cent in 1902 to 4 per cent in 1904. Without the best concentrator that had ever been built Bill could not have sold stock fast enough to make up the deficits.

All the time the mill was being built Bill was so far in debt that it looked as though he might lose the property any day. His Cananea stock was the only asset he had. The Wall Street wolves knew this, and every now and then they tried to run him out. In 1903 Gates and Hawley saw a chance to force Greene to sell for a song so that they could get his big interest for themselves. The fact that he had taken them in as partners two years before did not make any difference. They suddenly threw a lot of stock on the market and forced the price down ten dollars a share. If Bill could not get more security he was gone. Luckily he was used to protecting his own interests out on the frontier by any means that came in handy. He knew that one of his associates had over a million dollars' worth of Cananea stock in his desk. This was Bill's stock held under an option agreement, and Bill had to have it back. He went to the office, but they laughed at him. The laugh didn't last long. Bill pulled out a big six-shooter, and while his former friend cowered in the corner Bill broke open the desk, took the stock, and backed out.

He had plenty of security now. The next thing was to put the market price back where it belonged. He had a good play all ready to spring. A few months before one of the engineers of the Phelps Dodge Company had visited Cananea. The ore bodies were looking especially good just then. Bill took advantage of this to induce the Phelps Dodge to buy twenty thousand shares of Greene Consolidated stock. They bought it as an investment. It was a lot more than that to Bill. The Phelps Dodge Company was about the most respectable and conservative of all the copper companies. If the public knew that Phelps Dodge had bought Cananea stock at fifteen dollars a share, they would jump at the chance to buy it at six or eight. Just at the right minute Bill told the reporters that the Phelps Dodge Company was one of his biggest stockholders. Dr. Douglas, the blue-blooded president of Phelps Dodge, had to admit it. The Street had the story in a few minutes, and Cananea stock jumped up above par. Gates and Hawley had taken a good licking.

Tom Lawson tried his hand at freezing Bill Greene out in the market the next year. It was rumored that the Amalgamated officials were backing Lawson. They had turned Cananea down four years earlier, but now they wanted it. Lawson gave out reports that the Cananea mines weren't half as good as Greene claimed and that they were in desperate straits for ready money. At the same time he put in a flood of selling orders. The stock dropped like a shot. Bill was hanging over the edge of bankruptcy. Only a bold play could save him. He called in the reporters again and gave them a public letter to Tom Lawson. In it he called Lawson a crook and a liar and a cowardly cur, and said Lawson was trying to ruin him and his stockholders by false reports about the mines. Bill said he would go to Boston the next day – December 13, 1904 – to lick hell out of Lawson. Lawson answered, through the reporters, that Greene was a four-flusher and if he dared to come to his office, he would show him that his cheap Mexican bluff wouldn't go in Boston. Greene almost burst with rage. His private car was hitched on to the first train to Boston. All the way he gave the crowd of reporters harrowing and profane details of the way he was going to tear Lawson to pieces. Lawson sat in his big office and told the reporters how that "four-flushing —— — — ——" would have to get an ambulance to carry him home. The train at last pulled in. Bill took a taxi straight to Lawson's office. An eager crowd of newspapermen made it a procession. Lawson sat at his big desk like a volcano, spouting profane threats. Brushing the secretaries aside Greene stamped into the office. Lawson jumped up to his feet. With flashing eyes the enemies faced each other for a long silent minute. The reporters were ready to drop to the floor at the first move toward a hip pocket. The silence seemed to last an hour. At last Bill broke it. "Hell, let's have a drink," he said. Lawson said, "It's on me." They adjourned to the Parker House bar for a long one. Then they went back to Lawson's office and locked themselves in for four hours. What they said no one knows. They came out arm in arm, mellow with good whisky. The Lawson attack on Cananea was over.

With the big mill finished and Wall Street licked, Bill was on top of the wave. Every year Jim Kirk sent down more ore to the concentrator and the smelter – and every year Bill put more of his old Tombstone and Bisbee friends on the payroll. The price of copper was going up, but the Cananea costs soared still faster. The swarm of bookkeepers managed to show a paper profit that let Bill pay occasional dividends without going to jail. But to do it he had to charge two or three dollars to invested capital for every dollar of profit. By the middle of 1905 he had romped through twelve million dollars of the stockholders' money, and had paid them back less than three millions. Like a lot of people, he figured that the more men he had working for him, the more important he was. He wanted to be the biggest man in the Southwest, and plenty of people were glad to help him, at five dollars a day for doing nothing.

Tom Basset was the only one who got under Bill's hide about his management. Tom had built a sawmill across the valley in the Ajo mountains, and figured he could make a little money hauling mine timbers to Cananea. One day Bill was bringing in a bunch of New York directors to look things over, when he ran into Tom on the train coming down from Naco. Bill always liked to swell around a little in front of his old border friends, so he asked Tom to come back in the "Verde" to have a drink. The "Verde" was the finest private car ever built, and Bill had named it after himself — Verde being Mexican for green. Well, Tom came back, and after a drink or two Bill said to him, in a sort of superior way, "Why Tom, don't you know you could buy timbers for forty dollars a thousand up in Puget Sound and haul them down here by bull team cheaper than you can cut them in the Ajos?" And Tom answered "Yes, and you could buy copper in New York for thirty cents a pound and pack it out in this here dude car cheaper than you are making it in Cananea." Bill hadn't anything to say, but the directors pretty nearly busted wide open laughing. They knew it was costing Bill forty cents a pound to make copper out of 10 per cent ore.

The market was way up and money was coming easily now, so Bill could go after some of the other big projects that had been milling around in that imagination of his. He started in with a plan to get together all the old gold and silver mines in Mexico. He figured that with modern machinery he could make a lot of money out of low-grade ore that the natives couldn't do anything with. Of course, the Mexicans were delighted to sell their old mines to him at big prices. Between 1902 and 1905 he got together four thousand square miles of mining claims. Nobody had made any money out of a lot of them for fifty or a hundred years. They extended up and down the Sierra Madre for a thousand miles south of the border. Bill formed a new company called the Greene Gold and Silver to work these old mines, and listed the stock on the Boston Exchange. Before he got through he issued $3,000,000 worth of 8 per cent preferred stock and $22,000,000 worth of common. The brokers were sure it was going to be the biggest gold and silver company in the world. Bill put his old friend Ed Massey in charge of it. Before long he had thousands of men at work grading fine highways across mountains that a bald-headed eagle could hardly fly over without getting dizzy, putting up big hydro-electric power plants, and building cyanide mills that could handle five hundred or a thousand tons a day apiece. He even spent a little money underground. But Ed Massey was too prosperous to work much, and didn't go down in the hole any more than he had to now. His superintendents followed suit, and the Mexican miners made every shift a siesta. It would take many years to get the mines ready to supply the mills with five hundred or a thousand tons of ore a day at that rate.

The next big idea Bill had was to cut his own lumber and mine timbers. He was paying thirty or forty thousand dollars a month for Puget Sound lumber. Only a few hundred miles east of Cananea, in the Sierra Madre Mountains, there was a great forest of pines and firs. He ought to be able to start a lumber business that would supply most of Mexico and all the mines along the border. Bill's companies were selling far above par at that time, and he had no trouble raising a few more millions to finance the Sierra Madre Land and Lumber Company. He ran it in the usual Bill Greene style. Instead of a little logging road from the nearest railroad point on the border, he built a three-hundred-mile broad-gauge railway from El Paso to Chihuahua City by way of the timber concessions.

Great sawmills, with the most modern machinery, were to make all sizes and shapes of lumber at the rate of several trainloads a day. Hundreds of big Percheron horses soon killed themselves trying to work on the steep mountainsides that only sure-footed native cow ponies could negotiate safely. And an army of five-dollar-a-day Michigan teamsters and lumberjacks ate their heads off at the palatial company boardinghouses at the main camp of Madera. Mine timbers might about as well have been made of pure silver, as the labor cost was so high that the added cost of the metal wouldn't have made much difference. Bill kept the Sierra Madre Land and Lumber Company going by a twenty-year contract with his own mines, but no one else could afford to buy lumber from him.

Still his craving to be industrial Czar of the Border was not satisfied. Around Casa Grande, in southern Arizona, he had ridden over hundreds of thousands of acres of level desert. The large mesquite and palo verde trees proved that the soil was remarkably rich. Some vanished Indian tribe had changed this desert into a beautiful irrigated garden long before the white men came. Disease or a fiercer tribe drove them away centuries ago. Only the ruins of their many-storied "Big House" and dry grooves in the desert that were once irrigating ditches remained to tell of the ancient civilization. These relics of a forgotten race excited Bill's imagination. If the Indians had put water on the land, why couldn't he do it?

The source of the Indian's water supply was still within easy reach. The Gila River was only a few miles north of Casa Grande. Most of the year it was only a winding streak of gray sand, bordered by discouraged "batamote" reeds. But now and then a flood came down from the upper reaches, when the snows were melting in the Mogollons or a cloud burst hit anywhere along the Arizona-New Mexico line. Then the river boiled down in a torrent six feet deep and a half a mile wide. If Bill could build a dam to store the flood waters, he could irrigate all the Casa Grande Valley and sell fifty million dollars' worth of land.

His first dam-building effort, at his ranch on the San Pedro, ought to have discouraged him from monkeying with desert rivers. But the prospect was too alluring. Money came a little harder this time. However, a lot of people were willing to follow "Colonel Greene" in anything he undertook. He raised enough to tie up most of the valley land with Indian scrip and started to build a big earth dam across the Gila.

The Copper Company, the Gold and Silver Company, the Lumber Company and the Casa Grande Irrigation District would have been enough for most men to keep running all at once. But Bill was not yet satisfied.

One more project was too good to pass up. He had punched cattle so long that every time he saw the white faces of a bunch of Herefords peering around the turn in an arroyo he felt as though he was among friends. Now was his chance to show the world a real cattle ranch. All the other ventures he financed with other people's money. The ranch was all his own. Whenever the Copper Company declared a dividend Bill put his share of it into buying more land or more cows. He started with his little home ranch on the San Pedro, at Hereford. Gradually he bought mile after mile of range on both sides of the border. His pastures extended west over the high, timbered slopes of the Huachucas and across the grassy, oak-dotted knolls of the San Rafael Valley. This valley he made into a blooded stock ranch, where he raised the finest Hereford bulls in the country. South of the border he bought the Turkey Track, queen of all the cattle ranches in Sonora. With Cananea nearly at its center, this beautiful range extended for a hundred miles from east to west and sixty miles south from the border. The Sonora, the San Pedro, and the Santa Cruz Rivers all started in big springs around the base of the Cananea Mountains. Even in the driest years they brought water within reach of all the cows on the Turkey Track. When the rains failed and the grass on the broad mesas dried into brown dust his cattle could browse on the brush-covered slopes of four big mountain ranges. They still kept strong when the cows on neighboring ranches starved by the thousand. The Turkey Track herd grew to be the finest in the Southwest. Forty or sixty thousand they numbered — all big, white-faced Herefords that brought the highest prices wherever cattle were sold.

It was a principality, and Bill was its feudal lord. His vaqueros and their families filled half a dozen villages. When he made a trip of inspection over the railroad that ran through his range for a hundred miles, the Mexican foremen rode in from all parts of the ranch to pay their respects to the great "Meester Greene" who sat in state in the palatial "Verde" to receive them. They pretty nearly worshiped him. Bill never forgot to ask about their wives and children, and when the "Verde" was set out on a siding for a few hours, he wasn't too proud to invite them to sit down with him and eat a banquet such as they had never dreamed of. He was quiet and friendly, and yet he always made them remember that he was the big boss. All their questions and arguments he settled firmly and justly. No man ever knew better how to get along with the Mexicans. Dozens of them would have cheerfully committed murder for him if he said the word.

Through 1905 and 1906 Bill was on top of the world. In Cananea Jim Kirk had developed 40,000,000 tons of 4 per cent ore, and was running four thousand tons a day of it through the big mill and the smelter. The Copper Company paid nearly $2,000,000 in dividends in 1905. To be sure it had to raise $3,400,000 by selling stock the next year, but he said this was for working capital and for enlarging the plant still more. The Gold and Silver Company, the Lumber Company and the Irrigation District were in full swing. Ten thousand workmen jumped at Bill's orders, and twenty thousand women and children got their beans and tortillas from his payroll. No wonder Wall Street looked on Colonel Greene as the financial wizard of the Southwest. His stocks were all selling far above par. The market quotations showed a value of a hundred million dollars for the four companies. Half of the stock was in his own name. In seven years he had grown from a ragged cow man and adventurer to the greatest captain of industry the border had ever seen.

Fate had lifted Bill up on stilts. Before it knocked them out from under him, it made him the hero of a melodrama that was just the climax his romantic soul craved.

Bill himself always treated the Mexicans like white men. Unfortunately most of the other Americans in Cananea were not as fair. A lot of them thought the Mexicans were just dirty greasers. When they got drunk down in Ronquillo they didn't hesitate to say so, and they thought it was smart to shove the Mexicans off the sidewalks. Now and then a "gringo" would take a shot at a Mexican just to see if he was too drunk to shoot straight. On top of that the Americans held all the good jobs as foremen and bosses and timbermen. In many cases American miners got five dollars a day for doing the same work that Mexicans were doing for two dollars. And the Americans lived in fine plaster houses surrounded by gardens up on the Mesa, while the Mexicans were crowded together in little shacks built of tin cans or scrap lumber down in Ronquillo or on the mountainsides around the mines.

Naturally the Mexicans resented getting the worst of it in their own country. All the time the feeling grew more bitter. There were strikes every year or two. But the Mexicans had to eat, and a few days of hunger forced them back to work again. After a few futile attempts they were in just the right frame of mind to listen to the agitators who urged them to kill all the gringos and take over the mines for themselves.

The storm finally broke in May, 1906. A drunken American tried to take a girl away from a Mexican in a dance hall in Ronquillo. The Mexican reached for his knife, and fell with a bullet through his heart. Next morning two or three thousand Mexicans got together down in the gulch at the foot of town. They all had rifles or knives, and mescal had given them lots of nerve. For an hour or two they just milled around like a lot of cattle. Another drink started the leaders up toward the Mesa. The mob followed, shooting their guns in the air and yelling, "Muerto a los gringos." It was a regular stampede. The Americans who weren't at the mines got together at the far end of town, all ready to be massacred. Nothing could save them — except Bill Greene.

And Bill came through. The mob had reached the railroad station, at the edge of the American town. Bill drove out in front of them alone in his big open red automobile. Stopping it a few feet from the leaders, he stood up on the seat and held out his hands for quiet. The Mexicans kept on yelling and shooting. Bill never moved — just stood there, big and confident. Gradually the yells died down. Bill waited until there was a silence that was almost suffocating. Then he started to talk to them in Mexican, slowly and quietly, calling them "my friends." He told them they would be crazy to kill the gringos. That would just mean shutting down the mines, and they would starve. If there were injustices, he would take care of them. He had given them the jobs that fed their wives and children, and they could trust him. They listened, quiet but menacing. So he held them until Colonel Kosterlitzki rode in at full gallop with a company of Rurales. Don Porfirio was still in power then, and the Mexicans knew that when the Rurales started to shoot, a lot of them were going to get killed. So they slowly broke up and filtered back to their homes. The riot was over, and Bill Greene had saved his people.

The trouble came near breaking out worse than ever the same afternoon. Early in the morning frantic messages to Bisbee, across the border fifty miles away, had told of the desperate danger. The Bisbee miners and gamblers eagerly hurried to save their friends. All of them who could get rifles or shotguns or revolvers crowded on a train of flatcars and started for Cananea. Every man brought a couple of bottles of whisky. As they stopped in Naco on the way they loaded up with more of it. By the time the train reached Del Rio, six miles out from Cananea, they were the drunkenest lot of wild men that ever got together. They were all ready to shoot, and no one could tell which way the bullets would fly.

If they had reached town, it would have been a general slaughter of both Mexicans and Americans. Luckily the railroad men kept their heads and moved the train slowly. From Del Rio they wired ahead to warn Greene. So Bill and Governor Ysabel of Sonora and Colonel Kosterlitzki rode out and met the train a couple of miles from Cananea. They started it back to Naco before the invaders had time to climb off. Few of those on board were sober enough to know which way they were going until they were halfway across the valley. As they stumbled off the cars in Naco to get another drink, Bill Soule's gun went off and shot him in the foot. This was the only casualty in the famous Battle of Cananea.

Bill Greene was a hero now as well as a great industrial leader. But the cards were about ready to turn. Every man has just so much luck in his life, and Bill's luck had come all at once. When it changed, he fell clear to the bottom. He was still lucky, in a way, as he did not have to watch his fortune dribble away through heartbreaking year after year. If a collapse is sudden enough, it is almost as dramatic as a success. Bill loved anything theatrical, so he must have got at least a little satisfaction out of the way Fate wrecked him.

Cananea had started his big play, and Cananea ended it. Wall Street at last refused to buy more stock to cover the deficits. Debts crowded in on all sides. With the richest ore bodies in the Southwest, the Greene Consolidated was on the point of bankruptcy.

The Amalgamated Copper Company officials were waiting to take advantage of the emergency. For years they had coveted this rich plum. They already had secured a foothold in Cananea by buying the small America Mine. Their Cananea Central Copper Company, formed to work the America, had lots of cash but very little ore. Greene had all the ore in the world but could not meet his payrolls. Anyone could see what was going to happen. Bill tried desperately to fight his way out, but it was hopeless. Late in 1906 he had to surrender to his old enemies.

The Amalgamated crowd kept Bill's name, anyhow. They organized the Greene Cananea Copper Company to take over all the Cananea properties. Bill was down, and they drove a hard bargain with him. In return for their six millions in cash and the little America ore body they took control of the new company and of all of Cananea. Bill received a large block of stock, but he had to give it up again to make good the loans he had made from the Cananea treasury to his Lumber Company. In February, 1907, Cole and Ryan and their associates took over active control of the Cananea Consolidated. They sent down Dr. Ricketts to pull the mines out of the morass of extravagance. With his arrival, efficiency kicked romance out of the window. Bill still lived in his big house in the shade of the cottonwoods on the mesa and had the title of president of the operating company, but really he was hardly more than a pensioner.

The Sierra Madre Land and Lumber Company collapsed with the Copper Company. Before he accepted the job of manager at Cananea, Dr. Ricketts insisted that Greene cancel the twenty-year mine timber contract. This contract was all that had kept the Lumber Company alive. Bill had to sell out to the Pearsons for a few thousand dollars, that did not begin to pay the debts.

The Greene Gold and Silver Company lasted only a few months longer. The roads and mills and power plants were costing several times the estimated amount, and the treasury was empty. When the cards were coming his way, Bill could raise millions of dollars in a few days. Now he couldn't even get a few thousand pesos with which to pay the taxes. In July, 1907, the Greene Gold and Silver Company failed, with liabilities of $3,035,338 and $82.00 in cash.

The irrigation scheme was the only hope left. A cloudburst soon settled that. Fate wasn't taking any halfway measures to wreck the house of cards she had built up so carefully. The Gila rose and in ten minutes washed out Bill's earth dam so completely that hardly a trace of it was left. With money and credit exhausted, the Casa Grande Irrigation Company was as dead as a last year's snowstorm.

In less than a year the great Colonel Greene had dropped from the crest of the wave to ruin. Only the kindness of his old Amalgamated enemies kept him from bankruptcy. They let him keep his ranch for the four years left to him. He could still ride around over the broad grassy valleys and receive the homage of the Mexican ranch foremen who loyally honored their old friend and protector. But the glory had departed. His face grew sad and furrowed, and his gray eyes dimmed as they looked far beyond the blue mountains to the visions that had betrayed him. Fate was kind and did not keep him lingering too long with his memories. A fall from a carriage as he left his Cananea home ended the big game as he would have wished, with his boots on.

So Bill Greene joined the ranks of the empire-builders who fail to share in the proceeds of their building. Others had to bring his great plans to success. Yet he was the pioneer whose vision conquered the desert and the mountain. When History deals the final hand in the game of life, a lasting fame should fall to Bill Greene of Cananea.

AUTHOR'S NOTE: Colonel Greene's family told friends of mine that they bitterly resented my account of the early days of Bill Greene. If I came to Cananea, they said, I might be met with a rope. While I passed through Cananea several times, I had no chance to become acquainted either with the Greene family or with the rope. I cannot understand the protest by the family against what I thought was a sympathetic account of one of the most remarkable men in the history of the Southwest. My stories came from old Tombstone associates of Colonel Greene in early days. They at least thought they were telling the truth. But, as I had said in my foreword, there is probably no such thing as truth, especially in stories handed down by word of mouth. I think a man is all the greater because he came a long way from earlier troubles and failures. The story of Bill Greene's first forty years should make his descendants even prouder of his later accomplishments.

Part II of the present book will tell how large-scale open-pit mining is making up for the approaching exhaustion of the rich ore at Cananea. The prosperity that Bill Greene started will not be over for generations after his death.

The dark-colored hills were the outcrop of the richest part of Ajo's great New Cornelia ore body in 1915, before Calumet and Arizona started the open pit. Below, the fantastic McGahan Vacuum Smelter was still standing in 1912. *(—Both, photographs by the author)*

7

In a Mysterious Way—The Story of Ajo

TWENTY-SEVEN years on the road for a St. Louis dry-goods house is a record any man should be proud of. There wasn't a more respected salesman in all the St. Louis territory than John R. Boddie, and he knew it. He wasn't one of those cheap, smutty-story-telling drummers. He was a God-fearing Southern Gentleman, and when he said that a bolt of dress goods was all wool his customers didn't have to make any tests before they gave their orders. It did his friendly heart good to see the smile on the storekeeper's face when he came through the door and to settle down to a good three-hour talk on politics and the depravity of present-day morals and the iniquity of the Catholic Church.

John R. Boddie was the best single-hand talker in Missouri and Arkansas combined. Some of his competitors said the customers signed his orders because they were gasping for air under the flood of words and had to take desperate measures to get him out of the store. That was probably only jealousy, but anyhow John R. Boddie got the business and the southwestern territory was his just as long as he wanted it.

Still Boddie wasn't quite satisfied with life. Now and then he had to throttle a suspicion that there might be things more romantic than catching the ten-fifteen for Texarkana and selling a dozen assorted men's suits to Charlie Chamberlain. Still the Lord had been good to him, and if the Lord wanted him to have a more exciting lot, He could be relied on to take the necessary steps.

The Lord must have realized that here was one of His best supporters, who deserved a little extra recompense for all the years of unquestioning belief in the Methodist Church. It couldn't have been anyone except the Lord, for no one with less authority could have brought John R. Boddie the adventure and success he dreamed of through such an unpromising intermediary as A. J. Shotwell.

What Shotwell usually brought to those who were not wise enough to kick him out of the door was unqualified ruin. He was a fake mine promoter of the most venomous sort. For years he had been making an easy living off some copper claims in the abandoned camp of Ajo. Shotwell was shrewd enough to pick the ideal place for a stock swindler's mine. Ajo was in the heart of the Arizona desert, a hundred miles east of Yuma. Gila Bend, the nearest railroad station, was forty-five miles away. There was only one well on the track they called a road from Gila Bend to Ajo, and before the days of automobiles the drive across the glaring plain and the jagged lava fields was a desperately thirsty one. No stockholders were likely to come down to do any investigating unless they were pretty sure something was wrong.

Even if they did come, it was usually easy for Shotwell to send them home more enthusiastic than ever. The three low hills that rose in the center of the Ajo basin were perfectly adapted to catch the gullible investor. From a distance the porphyry that formed them was a beautiful golden-brown color, marred by no vegetation save the thorny arms of an occasional ocatilla or a parched gray creosote bush. But as soon as the rock was cracked, the fresh surface was brilliant green with copper carbonate. It was safe to let a stockholder hammer anywhere he wanted to in the fifty acres of the hills. Every piece he broke off looked like solid copper.

A little of it was really rich. Clear back in 1855, the hardy pioneers had hauled ore by bull team three hundred miles across the desert to San Diego and shipped it in sailing vessels to Swansea, Wales. There they sold it for $360 a ton. Unfortunately there were only a few streaks of this 50 per cent "ruby copper." Most of the ore that looked so rich was only stained by copper carbonate, so that it assayed one or two per cent copper. The first crop of prospectors soon failed, and left Ajo to the coyote and the Gila monster.

There was too much copper sticking out of the ground to remain long abandoned. A handful of the hardy "desert rats" who win a scanty livelihood in country where a jack rabbit would starve to death, drifted in one by one and located mining claims all over the Ajo Basin. Tom Childs and Rube Daniells ran a few bony cattle among the mesquites that lined the desert washes and managed to keep their Papago squaws and a rapidly growing crop of children alive while they did their assessment work on the copper claims that they hoped might some day be worth something. The rest of the group worked for Tom and Rube, or just hung on without any visible means of support. They all looked pretty tough in their faded blue overalls and torn cotton shirts. Frank Merrill didn't even wear any socks, and as he rode through the cactus after a cow, with his overalls up to his knees, his bare shins attracted all the thorns in sight. John Greenway asked him about it one day, and Frank answered, "I used to wear socks, but they was always getting ketched in the bresh and tore, so I quit 'em." Skin was cheaper than clothes out on the desert.

In spite of their tattered clothes and unrelenting poverty, Tom and Rube and the rest of them were real men. Their eyes were clear and steady from looking a hundred miles off across the desert, and there wasn't an ounce of quit or of crookedness in their make-up. There weren't any truer friends or more dangerous, straight-shooting enemies on the border.

It was a hard game they were up against late in the 'nineties, when Shotwell came in and took an option on most of the claims in the Basin. The low-grade carbonate ore was as worthless as the wild garlic that sprang up after the yearly shower and gave Ajo its Spanish name. The name was an appropriate one, for the odor just suited the sort of enterprise Shotwell planned.

When he went back to St. Louis to raise money to work the claims, Shotwell did not tell his prospective stockholders that they couldn't save the copper out of the carbonate ore. He called the hills the "Big Water Melon," and they were lucky to get a chance to help cut the melon. The mines were so rich he would not let any one company have more than one claim. If he did, the company might become too powerful and control the copper industry of the world. The St. Louis clothing merchants to whom he told the story were breathless with joy when Shotwell allowed them to buy stock in the St. Louis Copper Company, to finance a ten-stamp mill that was to treat the ore on the Cardinoff claim. John R. Boddie was one of them, and he told his customers that before many years he could retire and live on the dividends of his copper mine.

All the reports were rosy until the mill started to run. Then it developed that there was only water enough to keep it going half the time. The only water supply in Ajo was a slow seepage into the bottom of a few old shafts, and two or three hours' run even with the little ten-stamp mill dried up the shafts completely. Then the miners had to sit around and wait for the water to trickle in again. By picking out the richest stringers, Shotwell did manage to ship $36,000 worth of concentrates. But it cost him $45,000 to do it. The St. Louis Company went bankrupt, and the "Rescue Copper Company," which Shotwell formed to pull the St. Louis Company out of the hole, was evidently headed in the same direction.

Shotwell managed to get away with enough of the St. Louis and the Rescue Companies' money to live well and to make a few payments on his $200,000 option from Tom Childs and the other locators of the "Big Water Melon."

But the money was not coming in fast enough to suit him. Though he still planned to keep most of the claims for himself until he could get a big price for them, he was ready to let another company have part of the great ore body and share in the millions.

John R. Boddie was just the man to help form the new company. Shotwell told him he must keep it quiet, as the officials of Amalgamated and the other big copper producers must not know of this competitor. Boddie appreciated the compliment. He was indeed fortunate to be chosen to help the great Shotwell in this epoch-making enterprise. Behind closed doors Boddie explained the opportunity to the leading men in his territory. They knew he was honest, and anyone who could hold a job on the road selling dry-goods for twenty-seven years was certainly a safe guide in a mining enterprise. Captain Huie of Arkadelphia, Arkansas, W. W. Brown of Camden, Arkansas, and C. E. Neely of St. Louis were the bankers and lumber men chosen to organize the new copper company. In April, 1900, they joined Boddie on the train for a secret inspection trip to Ajo.

Shotwell was waiting for them in Gila Bend. After a good night's sleep in Captain Frank Welcome's livery stable they started on the forty-five-mile drive to Ajo in a Concord coach behind four good horses.

April was the ideal time for a desert ride. The brilliant sun was not yet high enough to bring the overpowering heat of the summer. In the early morning the distant mountains seemed sharply cut out of mauve cardboard, resting against the deep blue of the sky. Range after range, they rose for fifty or a hundred miles in every direction. As the sun rose higher, the mirage worked magic changes in the desolate landscape. Shimmering lakes came and went in the long slopes of the valleys. The black and gray lava peaks grew into pillars and battlements, changing moment by moment into a thousand strange geometric forms. The cactus and palo verde flung their tortured arms in weird angles against the gray creosote bushes of the desert. Now and then a kangaroo rat jumped across the road, or a coyote melted into the low brush like the ghost of a starved, mangy cur. As the sun sank behind the distant peaks, the gray of the mountains turned to pink and rose and violet, outlined against a sky of brilliant saffron. It was all so fantastic after the homely landscapes of Arkansas and Missouri that Boddie and his friends were almost dizzy with new experiences when at last they drove into the little basin in the Ajo hills. They spread their blankets on the board floor of the shack where Captain Hovey dispensed whisky and beans and dried meat to the Mexican and Papago Indian workmen, and dreamed of the millions of Croesus.

Shotwell had his plans carefully laid to make a good impression on the visitors. For weeks he had been gophering out rich streaks of ore and letting water accumulate in the old shafts. The pounding of the ten-stamp mill woke Boddie and his friends up at seven o'clock on the morning after their arrival. When they reached the mill after a greasy desert breakfast of beans and bacon, a band of high-grade concentrates six inches wide was coming off the tables. Samples that Boddie took of the ore pile assayed 5 per cent copper, and concentrates ran 45 per cent. Returns from the El Paso smelter showed that the concentrates were worth a hundred dollars a ton. Huie stayed to watch the mill, while Shotwell took the others over the brown hills. Every rock they broke was green with copper carbonate. Shotwell showed them how to hammer their samples of ore into powder and dissolve the copper carbonate in weak sulphuric acid. Then when they dipped their knife-blades into the solution, the steel was at once coated with pure copper. It was a spectacular test – and they didn't know that one per cent ore would give just as good a copper coating as 5 per cent ore.

Two days were enough to convince the clothing and lumber men that they had the richest mine in the world. They did not need any engineer to help them decide. They bought the four claims on the spot and paid Shotwell the $19,500 he asked, plus a large stock interest in the new company.

On their return to St. Louis they organized the Cornelia Copper Company. The name was in memory of Mr. Boddie's first wife – the only one he could think of worthy of the honor. Mr. Boddie explained this in touching words on a later visit to Ajo. "The name, suh, was that of the first Mrs. Boddie, the noblest and the loveliest woman that ever graced God's . . ."

At that moment the second Mrs. Boddie came around the corner and without a second's hesitation Boddie continued: "And in all my experience I have never known the flowers to be so beautiful."

The next thing was to raise money for the development of the new mine. There were a hundred thousand shares of ten-dollar stock authorized. As a special favor Boddie's old customers were allowed to have a little of it at $2.50 per share. Boddie himself sang the praises of Cornelia on all his rounds of the dry-goods trade. His former customer Charlie Chamberlain devoted his entire time to stock-selling. Dollar by dollar the money came in. Chamberlain perfected a plan that got results every time he tried it. He let the prospective stockholders in a new town select one of their number to visit the mine. Chamberlain went along and paid all expenses. After the visitor had cracked pretty green rocks until he was tired, Chamberlain followed the Shotwell assay method: powdered a little ore, dissolved it in acid, and let the "prospect" coat his knife-blades with copper. It never failed. As soon as the delegate got back home, he and all his friends bought all the Cornelia stock they could afford.

The venture proved to be more difficult than the promoters expected. A little development convinced Huie and Boddie that the Ajo ore was too lean to work with a stamp mill and concentrating tables, in a country where water was scarcer than whisky. Their enthusiasm was as great as ever, but after a few months of failure they had to shut down the Rescue mill and wait for the discovery of some new process for recovering the copper.

A company of lumber and clothing merchants looking for a new copper process was manna from heaven to Fred L. McGahan. McGahan was an ingratiating little Irishman with an explosive manner and eyes that were always darting from point to point like bright blue hummingbirds. He had all the technical terms of chemistry and metallurgy down pat. When he started to talk about how he could save the copper he just lulled his victims into a trance of greedy ecstasy. He had already swindled two St. Louis chemical companies with his fake processes. Boddie was too enthralled with the description of the McGahan Vacuum Smelter to do any investigating. Shotwell was just as enthusiastic. They signed a contract for all the North American rights to the new process just as soon as the papers could be drawn.

Boddie and Shotwell almost had a knock-down and drag-out fight to decide which company should have the honor of building the first furnace. Shotwell had organized the Shotwell Tri-Mountain Copper Company to open up some more of his claims. He wanted to be sure that he would not run out of stock to sell, so he made it a ten-million-share company. He kept six million of the shares for himself and started to sell the rest as fast as he could at twenty cents a share. Shotwell wanted the first McGahan furnace for the Tri-Mountain, as this would be a fine talking point in his stock-selling campaign. Boddie and Huie insisted that the Rescue or Cornelia should have first chance. Mr. McGahan said he could only superintend the erection of one furnace at a time, so two of the companies would have to wait. At last the Rescue and Cornelia companies joined hands and persuaded Shotwell that the first furnace should go on the Rescue. Cornelia and Tri-Mountain would get theirs as soon as McGahan could handle the extra work.

With the profits assured by this marvelous process, it was easy to sell stock in all three companies. Cornelia and Rescue were bringing in three to six dollars a share, and Tri-Mountain twenty or thirty cents. On paper, Shotwell and Boddie and the rest were already millionaires. The treasuries were soon overflowing, and McGahan ordered rush delivery on the steel for his furnace.

Thus far all had been happy and harmonious among the three companies. Only the squabble as to who should get the first smelter had thrown a momentary fly in the honey. But with prosperity, dissension began to spring up. Boddie and Huie became a little suspicious of the mining ability of Shotwell and appointed a new manager of their companies.

Shotwell got even by selling a lot of his own promotion stock in Rescue and Cornelia. This was treason, as the large sales put down the price of the stock and made it harder to finance the smelter. Soon the bosom friends were almost on shooting terms.

Shotwell had all the best of it for a while. With the money received for his Rescue and Cornelia stock he paid off all save $22,500 of the money he owed Tom Childs and the other prospectors for the claims. He had nearly $100,000 in the treasury of his Tri-Mountain Copper Company, and every cent the others spent helped him more than it did them. If the mines did make money, he owned control of the best of them, and if they failed, he had a fat bank account from his stock sales anyhow. Whichever way the luck broke, he was sitting pretty.

Just when he was on top of the world, Fate decided to buck him out of the saddle. "Small fleas have lesser fleas to bite them," and Shotwell had now sucked enough blood out of his stockholders to be worth plucking himself.

His only faith was his downfall. He had always been inclined to believe in spiritualism. At last he had money enough to test it. Whenever he went back to St. Louis to sell more stock, he hunted up a séance and tried to get some advance information on what the future held in store for him. At that time a forty-year-old medium named Mrs. Tharp was dishing out the secrets of the spirit world to those who had five dollars' worth of curiosity. Her husband, ten years her junior, gave eloquent lectures on the wonders of Mrs. Tharp's communication with the leading citizens of the land of shades. Between lectures he looked up the financial standing of any visitors who showed promise of being repeaters. Tharp soon spotted Shotwell. A little quiet investigation showed that here was the ideal spiritualistic meal ticket.

The beautiful girl from the spirit world who entered the unconscious form of Mrs. Tharp began to have messages for one in the circle who was destined to greatness. The messages were not yet clear. They seemed to concern a strange desert place, all the color of bright new copper. The name of the future Napoleon of industry seemed to begin with S. The rest was hazy. "Little Bright Eyes," as the spirit maiden was called, was only an intermediary, bringing communications from souls far too exalted to descend into Mrs. Tharp's plump form themselves. She must learn to interpret their desires a little at a time, lest she be blinded by their radiance. The forms faded, and she could say no more. Another time — perhaps.

Shotwell knew he was the one for whom the message was intended. His Destiny was in the hands of the greatest denizens of the world beyond. If he followed up this clue, no pinnacle of success would be too lofty for him to climb.

Mrs. Tharp and her husband realized the honor that was offered to them, to help the upward course of this new man of destiny. They consented to live in a house that Shotwell bought for them on Washington Avenue in St. Louis. A secretary named Thaw came along to record the proceedings for the benefit of science. Every evening there were séances, and new revelations. The public were admitted, at the usual price, but most of the messages were for Shotwell.

Little Bright Eyes grew ever more adept at bringing word from the souls beyond the grave. Montezuma was Shotwell's especial guide and prototype. Every night the King of the Aztecs sent advice to the one who was to reincarnate his greatness. After a time Shotwell could even ask questions of Little Bright Eyes, and a night or two later — after the necessary detective work by Tharp and Thaw — Montezuma would send back answers that conclusively proved his all-seeing wisdom.

Now and then the messages told where to sink a new shaft, or what town to try next in the Tri-Mountain stock-selling campaign. More often they were less concrete. Look out for Huie and Boddie, Montezuma said. They were plotting Shotwell's downfall. Nothing definite was yet in view, but he must keep in close touch with his exalted guide. There might be word of the greatest importance the next night.

Mrs. Tharp had learned by experience that it was dangerous to be too exact in the advice that emanated from her trances. A couple of years before she had lost one of her best supporters by telling him just how to run his tunnel on the Sunrise claim in the Huachucas. Turn it a little more to the right, the message said, then to the left for fifty feet, then sharply to the right, then to the right again. Every turn meant another twenty dollars to Mrs. Tharp, so before long the tunnel looked like a flattened corkscrew. It was just on the point of entering the copper bonanza the spirits promised when a round of shots broke out to the surface, only thirty feet from the starting point. The tunnel had made a complete circuit, and Mrs. Tharp had lost a customer. She was not going to make that mistake again.

Montezuma was not the only spirit who was eager to aid Shotwell. Others less exalted were equally ready to help. A tough old desert rat called Arizona Bob had been a friend of Shotwell's until too much whisky had sent him to join Montezuma and Little Bright Eyes some years before. Even in death, Arizona Bob had not forgotten the drinks Shotwell had given him. Now he could repay the debt. Every few nights he would ask – through Little Bright Eyes – if Shotwell had any enemy he would like to have killed. When Captain Huie was throwing Shotwell out as manager of the Rescue Company, Arizona Bob was rarin' to go. He wanted to drop Captain Huie down a prospect shaft right away. Shotwell wasn't quite ready to go that far, but he suggested that if Arizona Bob could arrange to put a curse on Huie so that he never could come back to Ajo, it would be greatly appreciated. And sure enough, Captain Huie planned dozens of trips to the mines, but something always happened to upset his plans. He never set foot in camp again in the ten years he lived.

The apparitions finally got on such friendly terms with Shotwell that they wanted him to have their pictures on his wall. Of course, this was the crown to all his happiness. A painter up in Chicago was attuned to the after world and could catch the likenesses. It was a terrible strain to paint from models so tenuous. Naturally the painter had to have a lot of money to make up for the exhaustion in which the effort left him.

What was money compared with such treasures? Before long Shotwell had a whole gallery of his guiding spirits. Montezuma, in all his kingly panoply, cost $3,000. A beautiful Egyptian princess of six thousand years ago who was waiting patiently to be Shotwell's bride beyond the grave set him back another $3,000, but she was well worth it. Before the occult gallery was complete, the artist and the Tharps, with whom he split the fees, were richer by $20,000.

Shotwell became so proud of his mystical friends that he took McGahan to a séance to meet them. This was poor judgment. McGahan's unstable mind was already so inflamed by cocaine or just plain insanity or whatever it was that ailed him, that he became an enthusiastic convert to Mrs. Tharp. She had a fine time playing one against the other. First Shotwell and then McGahan was invited to join the medium inside the cabinet that shielded her from the eyes of the less fortunate seekers after truth. There she made the lovely Egyptian princess appear most convincingly to each of them in turn. This was going a little too far, even if the soul-mate had been dead for sixty centuries. The two on whom the shadow world was showering its choicest favors became wildly jealous of each other. At last they were on the point of a real battle – which was a terrible state for men who were used to fighting with their wits and not their fists. McGahan gave in just in time to save his neck. He went back to Ajo, swearing that not a pound of ore from Shotwell's mine should ever go through his smelter.

Boddie and Chamberlain and the rest finally discovered that the money they turned over to Shotwell was going to Montezuma and the Egyptian princess and not to the very material Tom Childs, who was due the balance of the purchase price on his claims. They might themselves be liable to prosecution for obtaining money under false pretenses. To save their reputations and their companies they refused to have anything more to do with Shotwell and ousted him from all his offices. He told them to go to the devil, for the claims were in his name. But the Tharps were too grasping to let him keep the few thousand dollars he needed to complete his payments. The option was forfeited and the deeds released from escrow.

Tom Childs realized that the stockholders were not responsible for the default. The Tri-Mountain and Cornelia and Rescue companies had for the most part the same stockholders. Therefore Tom and his associates sold the claims to the Cornelia and Rescue companies for just the amount of the balance due under the original option. These two companies then consolidated, bringing all three Ajo hills under the ownership of the Cornelia Copper Company.

Shotwell claimed fraud and brought suit for $200,000. The spiritualists had cleaned him so thoroughly that he could not raise money enough to furnish a bond for the costs. The suit was thrown out of court without a hearing, and Shotwell disappeared from the story of Ajo.

All the time Shotwell was communicating with the spirits, McGahan was busily constructing his vacuum furnace. It was a marvelous piece of equipment. The furnace itself was a brick-lined steel cylinder twenty-five feet high and six feet in diameter. Supported on a steel frame were smaller horizontal cylinders, to hold the oxygen and hydrogen gases that entered into the reactions. There was a powerful pump to maintain the vacuum. A bewildering array of pipes, gauges and spigots stuck out from all sides of the furnace. McGahan's marvelous discovery was that when the air was pumped out of the furnace and the ore fed in with a little fuel oil and just enough oxygen to burn it, he could regulate the temperature so accurately that all the elements in the ore would be melted, one by one. First the gold would melt and sink to the bottom, where it could be drawn off through the lowest spigot. The next spigot was for silver, and the next for copper, then one for calcium, then sodium and silicon and finally up at the top, spigots for oxygen and hydrogen gas. After the furnace was once started, he could burn the hydrogen again with the oxygen, and so get along without any outside fuel at all. Nothing was lost, and all the elements came out absolutely pure. John R. Boddie and his friends were the luckiest men in the world to have control of so matchless an invention.

The Ajo furnace was nearly completed when an opportunity was presented to Boddie and Huie that was far too good to pass up. While they were in Los Angeles to superintend the shipment of equipment to Ajo they were fortunate enough to meet the directors of the Gold of Ophir Mining Company. This company owned a spectacularly rich gold vein in the Calico Mountains, out in the California desert east of Barstow. The ore, the directors said, averaged $5,000 in gold per ton. Hundreds of thousands of tons of it were all ready to mine. The only difficulty was that the gold occurred in a rare telluride, from which the ordinary metallurgical processes were quite unable to recover it. Even the regular assayers could not find this gold when they tested samples from the Gold of Ophir. Often they would report only fifty cents a ton when the ore actually contained five thousand dollars. Luckily one assayer in Los Angeles had discovered how to make a correct determination. Otherwise Boddie's new friends might never have known what a bonanza they owned.

Boddie could sympathize with their plight. He had been just as unable to recover the Ajo copper until McGahan had come along to solve the difficulties. Why could not the McGahan Vacuum Smelter save the Gold of Ophir as it had the Cornelia? The gold could not elude this process no matter if it was in tellurides. Huie and Boddie soon convinced the officers of the Gold of Ophir Company that in justice to their stockholders they must put in a McGahan plant. Sad to say the Gold of Ophir treasury was empty. Huie and Boddie could not let this inconvenience block so great an opportunity. They agreed to build a little three-ton plant in Los Angeles with Cornelia money, in return for a few thousand shares of Gold of Ophir stock. McGahan was prevailed on to assume this extra responsibility, and the new smelter was rushed to completion. Gold of Ophir stock of course boomed, and its officers sold enough of it to amply repay them for their efforts.

The two smelters were ready to start at the same time. After some hesitation the Cornelia directors decided to blow in the one in Los Angeles first. McGahan could tune that furnace up for a few days, until it had produced two or three hundred thousand dollars' worth of gold. Then he could start the plant at Ajo. Huie, Boddie, Dr. Wallace, Chamberlain, and several other prominent stockholders assembled in Los Angeles to witness the great event that was to lift them into affluence.

McGahan made a last inspection and gave the word that all was ready. To Boddie went the well-deserved honor of lighting the first fire in the new furnace. As he did so, the stockholders shook hands happily and wired their friends that the end of all of Cornelia's troubles had come.

McGahan explained that a slow fire must be kept up for twenty-four hours, in order to prevent cracking the steel shell. By nine the next morning it would be safe to charge in the first ore. The stockholders agreed to meet again at that hour and separated to dream of the fruition of all their hopes.

At the appointed time they came again to the new smelter. The oil fire was still burning cheerfully. But no McGahan was in sight. They waited in the galvanized iron shack that hid the furnace from prying eyes, and watched the glow of the fire through the open vent near the top of the cylinder. An hour passed, and then another. Even Boddie's inexhaustible flow of words was damned by the foreboding that clutched at his vitals. They stood in anxious silence broken only by the sputtering of the oil fire. Still no McGahan. Could he have met with an accident, or become suddenly sick? At last they returned to the hotel. And there they found a letter from McGahan, saying that the $34,000 and expenses that they had paid him was far too little for his great invention. If they would pay him $50,000 more, together with a lot of Cornelia stock, he would return and blow in the smelters. Otherwise he could laugh at them, as no mortal save himself had any idea how to run the plants.

The awful truth burst on Boddie and his friends. They had been fools and dupes, not once but three times over. They had thought the whole world was honest and simple-minded like themselves. And their faith had been rewarded by the loss of their money and their friends' money. Their reputations were wrecked, and they were a laughing-stock and a mockery. Shotwell had fooled them, and then McGahan had made suckers of them again, and finally McGahan and the Gold of Ophir fakers had caught them a third time with the same bait. The best they could hope for was to have their stockholders call them fools. They would be lucky if they were not prosecuted as rascals to boot.

They could at least stop further expense at Ajo. So they sent Dr. Wallace to discharge all the workmen. McGahan's nephew Flagler was superintendent, and a friend of Boddie's named Rumph was clerk. Dr. Wallace told Rumph to "shut her down." Rumph found Flagler unscrewing the spigots and oil valves of the furnace, as though to prevent its use. Rumph told him to stop. Flagler cursed at him and said he would shoot him if he did not keep away. Rumph replied by putting three revolver shots through Flagler's head and chest. Dr. Wallace relented and kept Flagler alive. The sheriff drove over from Tucson to arrest Rumph, but after an investigation just called him a damned fool for not shooting straight and let him go.

The Cornelia Company tried to get back some of the money they had paid McGahan by suing him for obtaining money under false pretenses. McGahan replied with a $200,000 counter-suit for damages. Experts sent to examine the smelters said that all they could do was to swear McGahan was crazy. The Cornelia directors regretfully decided there was no use trying to get the best of a crazy man in court and agreed to cancel all the suits.

Boddie and his associates were in the depths of despair. They had used up all their money, and the Ajo ore was just as impossible to treat at a profit as it had been in the days of Shotwell's ten-stamp mill. Still they kept their faith in the property. All that copper must become valuable some day. They had to abandon Ajo once more to the handful of desert rats who were too poor to move away. But they managed to pay the taxes and held on in hope that some day their luck might turn.

In 1907, it looked as though their faith was going to be justified. Jackling's success with the Utah Copper had put an entirely new aspect on low-grade copper deposits. The desert was alive with engineers hunting for new "porphyries." It did not take them long to find the brown hills and spectacular copper stain in Ajo. Here was just the place to drill for a great disseminated copper ore body. The big copper companies were actually bidding against each other for a chance to develop Boddie's folly. The General Development Company, with Mr. J. Park Channing at its head backed by the Lewisohn interests, finally secured an option on a majority of the stock of the reorganized New Cornelia Copper Company. Mr. Seeley W. Mudd and associates, who had been largely responsible for developing the great Ray Consolidated Copper Company, optioned the Randall Ore Reduction Company, which owned claims on the south edge of the Ajo Basin. A group of English capitalists took an option from Tom Childs on some outlying claims in the basin itself, east of the Cornelia hills. All three groups started development at the same time and Ajo was humming with activity and hope.

Science had come to take the place of the charlatan and the ignorant promoter. John R. Boddie felt that the sun was shining on him. But Science was not yet sure of her ground, at least as far as copper deposits were concerned. The engineers of the three new companies were among the greatest in the mining world. They had all enjoyed spectacular success in other districts. In the course of their successes they naturally acquired theories that guided the development they planned. Strange to say all three theories were different. And stranger still, all three were wrong.

The engineer for the English syndicate belonged to an earlier German-trained school. He had been taught that ore bodies were associated with the heavy, dark-colored "basic" rocks, and his European experience had borne this out. The rocks were darker and more basic in Tom Childs' claim than elsewhere in the Ajo District. Therefore the big ore body must be under the bottom of the basin east of the brown hills. He started a shaft to prove it.

The engineers for the Mudd-Wiseman syndicate agreed with the more recent American theory that ore bodies are associated with the light-colored "acid" intrusions. The Randall property was largely covered by a nice white iron-stained rhyolite, one of the most acid of rocks. That was the place where they figured the ore body should be, and they brought in a churn drill to block out the ore as they had done at Ray.

The General Development Company engineers did not care so much about the exact character of the rock. But they were sure it must be soft and altered by the oxidizing action of surface waters in order to allow the leaching out of the copper near the outcrop and its concentration or "enrichment" in depth to form ore bodies of commercial grade. The three hills on the New Cornelia property contained a fair amount of copper, they realized, but the rock was far too hard to allow the necessary enrichment. Surrounding the hills they found a broad band of soft, highly altered iron-stained rock that was just what they were looking for. They brought in a diamond drill and started a circle of holes around the base of the hills.

In spite of the convincing theories and many thousand dollars spent in development, all three of them missed the ore body. The Mudd-Wiseman syndicate found a few rich stringers but no sign of the great disseminated ore body they hoped for. So they gave up their option and were the first to quit. The English syndicate did not find any copper at all, so they went back home at about the same time. The General Development Company did find fairly good ore in one drill hole that was right at the base of the hills. The other four holes were barren. They called in the greatest engineers in the country to make sure they were right. John Hays Hammond, A. Chester Beatty, and J. Park Channing all joined in the verdict. There was no ore in Ajo except for a shallow deposit of copper carbonate in the hills, and that ore could not be treated at a profit by any known or probable process. The General Development Company in turn gave up its New Cornelia option and abandoned Ajo.

Science had turned its back on John R. Boddie and left him in a worse hole than that dug by Shotwell and McGahan. What chance did he have now to win back the money he had lost for his clothing and lumber friends? They never wanted to hear of Cornelia again. Boddie might keep the company alive if he wanted to but he would never get another nickel from them.

Still Boddie stood by his childlike faith that there was a great ore body in Ajo. All the engineers in the world could not budge him from that belief. For a while he kept Charlie Chamberlain's son Lee at Ajo trying to run the little stamp mill, but could not quite make it pay expenses. Another chemical sharper like McGahan relieved Boddie and a few of his stockholders of $20,000 more for a hydrofluoric-acid leaching process almost as fantastic as the vacuum smelter. The leaching plant did turn out a few pounds of copper, but it cost a dollar a pound. The New Cornelia had to shut down tight once more and hope that some day something would turn up.

Even Fate must eventually weary of kicking a man who keeps on coming back for more. So when Fate couldn't kill John R. Boddie's faith in Ajo, it finally relented and proved he had been right all the time.

In the spring of 1909, when the Mudd-Wiseman syndicate and the General Development Company were finishing their drilling, a young geologist for the Calumet and Arizona Mining Company* happened to pass through Ajo on his way to examine a prospect of Jeff Milton's down on the border. Jeff had brought a bottle of whisky for his old friend Rube Daniells, and of course he had to wait a day in Ajo to help drink it. The geologist was glad to put in the time looking around the

*The young geologist was Ira Joralemon.

John C. Greenway had made New Cornelia a great success by the time this picture was taken at Bisbee in 1921. *(—Photograph by the author)*

Ajo Basin and getting acquainted with the men who were doing the drilling. He was greatly interested and a little amused by the three rival theories, though all the other engineers were so much more experienced that he hardly dared acknowledge his amusement even to himself. Any or all of the theories might be right for all he knew. But he did think that if he were doing the work, he would put one hole down in the shiny brown hills. In spite of the theories it would be interesting to see what lay under all that copper stain.

After regretfully deciding that Jeff's mine was too far from the railroad to pay, the geologist went back to Bisbee and forgot all about Ajo.

Two years later a new general manager came to the Calumet and Arizona. The engineering staff were greatly impressed and a little scared. Even at thirty-nine years of age John Greenway had made a great name for himself at Yale, in the Rough Riders, and on the Western Mesabi iron range. He had just finished successfully mining and milling at a big profit low-grade iron ore that everyone thought was worthless, and he was looking for other tough problems to solve. It didn't take more than one look at his straight lips and square chin to know that he was going to be the Boss.

So when John Greenway told the geologist that he wanted a steam-shovel copper mine, something had to be done about it. A search through all the old reports in the company files was fruitless. But the name of Jeff Milton's mine brought back a picture of the Ajo hills. What was under them, anyhow? It was a desperate chance after the best engineers in the country had turned the camp down, but it was the only hope. John Greenway wanted a steam-shovel copper mine, and if the geologist could not find one, his job was gone. Anyhow the chance was worth another trip to Ajo.

Lee Chamberlain took the geologist out from Gila this time in a decrepit Tourist automobile. Lee was trying to make a living in Tucson, running a rental car. His idea of economical automobiling was to take along just enough gasoline to get him back home. Half the time his patrons had to walk back.

Anyhow it was better than the trip the geologist had made two years before. Then it had taken three days in a freight wagon drawn by Sam Clarke's ill-assorted team, that consisted of a ratty little mule and an even smaller flea-bitten roan mare, led by an Indian stallion no bigger than a good jack burro. Lee's wreck of a car was palatial by comparison.

Ajo was at its lowest point then, in February of 1911. There were only four white men in camp – Tom Childs and Rube Daniells, Sam Clarke and Levy. Tom Childs and Rube Daniells were looking after their cows and half-breed children, and Sam Clarke had a lease on the Randall property and was trying to sort out enough rich ore to keep himself and two or three Mexicans from starving. Levy was a refined, quiet-speaking French Jew who had wandered up from Mexico years before and opened a little store. What brought him to that far corner of the desert no one knew. He must have come from a cultured family in Paris. But he stayed on among the rough pioneers and the Mexicans and Papagoes, making a bare living in his little store, the loneliest soul on all the border.

A few days of study and sampling convinced the geologist that there might be a great mine at Ajo in spite of the failure of the earlier development. A short tunnel in one of the brown hills had developed uniform carbonate ore running 2 per cent copper. Two or three shallow shafts had found sulphide ore of about the same grade. It was very different from any of the "porphyry" ore bodies that the Jackling companies had developed. Apparently the copper near the surface had been oxidized in place to a carbonate, instead of being changed to a soluble sulphate and leached out as at Bingham Cañon or Ray or Nevada. And the sulphide below it was not a soft enriched chalcocite ore, but was intensely hard, with minute specks of copper sulphides scattered through the rock. The geologist estimated that there ought to be ten million tons of the carbonate ore, and much more than that tonnage of sulphide. The carbonate could not be recovered by any existing process, and it was doubtful if the flinty sulphide ore, with the little specks of chalcopyrite and bornite, could be made to yield a profit. It was a tough problem. But John Greenway had been up against tough problems before and had solved them. With some hesitation the geologist recommended a few drill holes to prove that the twenty or thirty million tons of ore were really there.

Greenway agreed that it was worth a try. So one day John R. Boddie came home from his regular spring trip for the St. Louis clothing house for which he was still traveling to find a letter asking him on what terms he would option the New Cornelia property to the Calumet and Arizona. Boddie was so happy he almost had apoplexy, for Cornelia seemed dead beyond reviving just then. But he kept his head and drove a hard bargain. It was six months before Greenway could get him to the point of signing an option for 70 per cent of the 1,200,000 authorized shares of New Cornelia stock at about $1.60 a share, with the provision that the stock should be issued as the money was spent for development and equipment. The other 30 per cent Boddie kept to repay his stockholders for their many years of waiting.

The winter of 1911 was well along before the Calumet and Arizona started to drill the Ajo hills. Everything went wrong at first. The diamond drill runners were not used to such a hard, shattered rock. The first drill hole started was in a rich spot, but the bit became stuck a few feet down and this hole was not completed until three others had been finished. The second and fourth holes were on the edge of the ore body and were lean. The third hole was in the center of the richest ore, but it happened to hit a lean band that only assayed 1 per cent copper. Chances looked pretty slim when these holes were finished.

To make it worse, a $50,000 payment was due on a $200,000 option on the old Randall Group, that the Mudd-Wiseman syndicate had drilled two years before. It was out of the question to make this payment, when the first three holes completed on the adjoining claim were so lean. The Randall company would give no extension to the option. Finally Greenway had to be contented with a verbal promise that he would be given first chance at another option in case someone else made an offer.

Then one after another the succeeding drillholes on the Cornelia property developed rich ore. James Phillips, who had made many million dollars by optioning ground adjoining the big mines at Bingham and Ely, got wind of the Ajo development. He tried to option the Randall, but the Boston owners stuck by their word and insisted they would give Greenway first chance. Phillips was too much of a lawyer to be stumped by that. He persuaded the Randall people that an outright sale was not an option. He paid them $200,000 in cash and became full owner of the seven claims. He spent a million dollars more in development, and then sold the property to the New Cornelia for cash and stock worth nearly four millions. The hard luck with the first drillholes cost Calumet and Arizona and its New Cornelia subsidiary three million dollars.

The bad start did not discourage John Greenway. He was going to see it through in spite of the fact that the distinguished engineers who had condemned Ajo two years before urged him to abandon the work to save his reputation. After the first three holes it was easy sailing as far as the ore body was concerned. Within two years twenty-five thousand feet of drilling had proved that the Ajo hills were underlain by 30,000,000 tons of 1.5 per cent ore. Thousands of feet of shafts were sunk to make sure that drill returns were accurate. The work was so carefully planned that when the hills were finally mined, the grade of the ore agreed exactly with the estimate. John R. Boddie's reiterated statement that "where there are bear tracks there is bound to be a bear" was at last proven to be correct.

Finding the ore was only the first step in Greenway's Ajo campaign. Even if the copper was there, no one knew how to get it out of these new types of ore at a profit. Greenway was sure that if there was ore enough to justify the experimental work, he could find out how to treat it.

As soon as half a dozen drill holes had developed ore of satisfactory grade he set to work. It was a job that demanded the best engineers and metallurgists in the country. This meant, of course, Dr. L. D. Ricketts. Greenway employed the Doctor as consulting engineer, and the two of them were about the best team that ever tackled an engineering problem. Greenway's energy, practical sense and genius for administration were a perfect foil for Dr. Ricketts' technical skill. Both of them had the problem on their minds day and night for three years. Dozens of chemists and metallurgists helped to work out the new process. Every suggestion, however foolish it sounded, was thoroughly tested in the laboratory. As soon as they developed a method that looked good in the laboratory, they built a miniature plant at the Calumet and Arizona smelter in Douglas, Arizona. This plant ran for several months, treating one ton a day of ore that was hauled out from Ajo. The process was changed a dozen times to meet unexpected difficulties. At last the one-ton experiments were running smoothly. This was not enough. The next step was to build a "pilot plant" that would treat forty tons of ore a day at Ajo. New troubles developed and were corrected. After six months the forty-ton plant was running satisfactorily. Greenway and Ricketts had spent hundreds of thousands of dollars in the experiments, but at last they knew they were right. They reported to the Calumet and Arizona directors that they had developed a process of leaching with sulphuric acid that would produce copper from the New Cornelia carbonate ore for eight and a half cents a pound. As the average selling price was fourteen cents, this would yield an excellent profit.

One more problem remained, that seemed almost beyond hope of solving. The new leaching plant, to treat five thousand tons of ore a day, would require a small river of fresh water. Water was so scarce in the desert around Ajo that the few dozen men who worked on the drilling and the metallurgical tests had to drink stagnant rain water, caught on the galvanized iron roofs of the houses and tents during the rare thunder storms and stored in tanks for eight or ten months. The shafts sunk to test the drilling results made only a few gallons of water an hour. The nearest real flow was in the Gila

River, fifty miles away. That was a possible source, but a staggeringly expensive one.

There was just a chance that under the lava beds that covered the desert there might be a buried layer of gravel that would carry a fair supply of water. It was worth trying, anyhow. Greenway authorized a deep drill hole in the desert valley six miles north of Ajo. A great oil-well drilling rig was hauled out from Gila by mule team and set up among the dusty creosote bushes. Water for drilling had to be hauled from the trickling supply in the shallow well at Tom Childs' ranch. Down went the drill hole – a hundred, five hundred, a thousand feet. The lava was like steel. Sometimes it took days to make a foot of progress. The drill hole cost more than a hundred thousand dollars. But at twelve hundred and fifty feet depth it broke into a loose gravel that was full of water under a head that lifted it six hundred and fifty feet up in the hole.

Pumping from the drill hole failed to lower the level at which the water stood. Then came another hundred thousand dollars' expense for sinking a shaft and installing big electric pumps just above the level to which the water rose. The pumps lowered it only a few inches. The water was hot from the uncooled lava through which it flowed, but was remarkably pure. The supply proved to be inexhaustible. The last serious problem at Ajo had been solved.

For the next two years Ajo was alive with big construction. Greenway brought Mike Curley down from Minnesota to run the new operation. Curley had been superintendent of the Hill Mine for the Steel Corporation under Greenway and knew all there was to know about steam-shovel operations. It was a big change from Minnesota to the heart of the Arizona desert, but Curley soon became as much at home as the most hardened desert rat. A fifty-mile railroad from Gila was running eight months after the first dirt was turned. In fourteen months more the five-thousand-ton leaching plant was completed. Meanwhile Curley had built a beautiful Spanish town to house the workmen, with oleander trees shading the lawn that filled the plaza. In spite of the terrific summer heat, Ajo became a comfortable place to live in. The transformation from unbroken desert to a busy town of eight thousand people, with a great leaching plant covering hundreds of acres, was so rapid that it seemed like the work of a magician.

In April, 1917, the big plant was ready to run. It was an anxious time for Greenway and Ricketts and all the others who had worked so hard to plan and build it. They had tried to foresee every possible difficulty. But a new process is always a dangerous thing. The Chile Copper Company, in South America, had started a leaching plant along different lines two years before, and it was not yet working satisfactorily. If those in charge of New Cornelia had made any slip, it might mean the loss of $6,000,000 of Calumet and Arizona money. And with the money would go the reputations of a lot of good men.

There was no need to worry. The leaching plant worked perfectly from the start. Greenway and Ricketts, Curley and

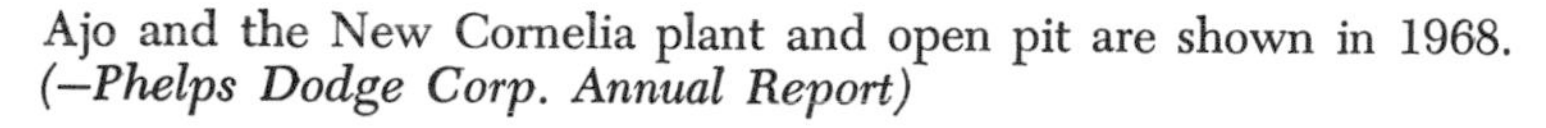

Ajo and the New Cornelia plant and open pit are shown in 1968. *(–Phelps Dodge Corp. Annual Report)*

the others had figured correctly at every point. The ore yielded exactly the estimated grade; the leaching plant obtained the estimated recovery; and the cost was within the estimate in the first month. The whole New Cornelia operation was a magnificent triumph of engineering skill.

John R. Boddie was overwhelmed with joy and admiration. All his dreams were more than fulfilled. His old friends of the clothing and lumber business who had followed his advice so trustingly were soon receiving dividends of 50 or 100 per cent a year on the cost of their Cornelia stock. Boddie himself no longer had to travel around selling clothing and listening to the gibes of disappointed stockholders. He was independently wealthy now, and he had made his friends rich with him.

After all the fraud and insane folly, Fate and John C. Greenway had made Ajo one of the greatest mines in the world. It is still growing. The ore body now measures a hundred and fifty million tons instead of thirty. Before the history of the camp is finished, the dividends will reach nearly a hundred million dollars.

The vision that Shotwell planted in Boddie's credulous mind had wandered through a strange and devious path. The swindling stock salesman, the fake spiritualist, scientific fraud and scientific error had done their best to kill it. But it kept on popping up again, inviting someone else to have a try at it. Ideas apparently have to wait until Fate is ready for them before anyone can make them practical. Then nothing can stop their fruition. John C. Greenway and John R. Boddie lived in two completely different worlds. When Fate was ready for another copper mine, it brought them together, and the result was Ajo.

AUTHOR'S NOTE: The New Cornelia operation has turned out to be even greater than I hoped when *Romantic Copper* was written. The chapter on the great low-grade mines in Part II will tell how the successful mining and treatment of ore that assays only a quarter of a per cent copper has increased the total tonnage that will be produced at Ajo to several hundred million and profits should approach half a billion dollars.

8

Fashions in Mines

WHEN the world embarks on a new industrial age, Fate and Science must work overtime to avoid economic chaos. The flow of materials and the direction of labor are rapidly changed. Men who have learned to vary their institutions a little at a time, through several generations, must make far more radical adjustments within a very few years in order to keep the new age from going on the rocks.

The Age of Electricity brought up some especially difficult problems. One of the most vital of them was the supply of new copper mines at the right time to meet the rapidly multiplying demand. The generation and use of electricity depend absolutely on copper. As more and more electrical machinery was installed, the consumption of this metal doubled every few years. The result was that in 1929 the world used twenty-one times as much copper as it did in 1860.

The mines that had met all the requirements of ten thousand years — the Cornish mines, Rio Tinto, and a handful of others — were completely snowed under by the demand. If a lot of new ore bodies had not appeared at just about the right time a copper famine would have made the rapid spread of electricity impossible. And if the new mines had been found too soon, overproduction would have soon reduced all the copper miners to bankruptcy.

Looking backward, the problem looks like a hopeless one. There are many districts so great that their premature discovery would have wrecked the copper industry. If Butte, for instance, had been developed twenty years earlier, it would have put the metal price down so rapidly that all the copper miners would have been ruined for a lifetime. Meanwhile no one would have hunted for new ore bodies. When growing consumption at last caught up with the excess productive capacity, no mines would have been ready to supply the demand. An acute copper shortage would have resulted.

Strange to say, in the half-century since the advent of electricity made copper so indispensable for our growing industrial civilization, new copper districts have been discovered just when the world needed them. To be sure, the development of a lot of mines within a short period has often led to overproductive capacity for a few years. But the excess production has never been acute, compared with the average growth of copper consumption.

This seems a wild statement to make at a time when nearly all the copper mines in the country are shut down or struggling along at a fifth of capacity; when the stock of unsold copper is decreasing at a hardly perceptible rate; and when the price of copper is so low that even the lowest-cost mines are giving away their best ore for a song. Yet it is true. If the use of copper had grown even as gradually as it did before the World War, the new mines would have come just in time to supply the required production without excessive cost. It was only the insanely violent fluctuation of industry in the past few years, doubling the consumption of copper between 1914 and 1929 and then cutting it in three between 1929 and 1932, that led first to the foolishly rapid equipment of new mines, and then to the stagnation of the whole copper business. From the point of view of today the discoveries of the past eight years have been disastrous. From the longer range point of view of fifty years from now, with the little wiggles in the curve of progress ironed out by Time, the discovery of all the new copper mines came just when the added supply was most needed.

Fate devised a simple plan for regulating the finding of these new mines. Different as they are in detail, nearly all the great copper districts can be grouped in four or five general geological classes, within which there are many resemblances. By some quirk of chance, districts of different sorts were not found indiscriminately. Instead, nearly all the deposits of one type were discovered one after another, within a very few years. Then another type came to the fore, and all the new copper ore bodies followed this new pattern for a while. It almost seems as though the Destiny that guides prospectors had put on a series of fashion shows, dressing up the copper mines of any one period in the same geologic style.

Of course, progress in the art of mining and metallurgy had much to do with the fashions in mines. When a new method of ore treatment made valuable the low-grade disseminated copper deposits, in which copper minerals are scattered sparsely through tens of millions of tons of porphyry or other rocks, prospectors and engineers started to hunt vigorously for these porphyry coppers. They soon found all the great ones. The opening up of a new country has often resulted in developing a number of similar mines. Yet this is by no means the whole story. Fate has often held back the discovery of rich deposits that would have been valuable at any time until bodies of that particular sort came in style and until the copper was needed.

Fashions in copper mines really started in the Copper Country of Michigan, in the 'fifties and 'sixties of the past century. At that time the growth of consumption was slow compared with that in the following decades. But it was enormous compared with the amount of copper then used. Copper was more or less a luxury in those days, used for pots and pans and roofs of large buildings. Machinery that required copper was almost unheard of. If a really great copper mine had been discovered in the early 'fifties there would have been no possible use for the copper.

Luckily the Lake Superior prospectors were stupid or unlucky enough to take twenty years to find the ore bodies that lay under their feet. By finding first the little "Mass copper" veins, then the much greater deposits in lava beds that contain almond-shaped or amygdaloidal gas cavities, and finally the really great Calumet and Hecla Conglomerate vein, replacing a bed of pudding-stone or conglomerate many thousand feet long, they carried the United States through the violently fluctuating demand of the Civil War and the following ten years without any very serious scarcity or oversupply. A rapidly growing demand caused the price of copper to range from twenty to fifty cents a pound between 1850 and 1875. Still no mines were developed large enough to glut the market. These twenty-five years were halcyon days for the copper miners.

Even the Calumet and Hecla Conglomerate could hardly supply the demand in the 'seventies. For the use of copper continued to grow in spite of the depression of 1873 to '78. Another great high-grade district was needed to bring the copper price down to a point that would encourage the adoption of electrical machinery. At just the right time, the Butte veins were discovered. These were tabular deposits of copper and iron sulphides with quartz, standing at steep angles. They were continuously mineralized for thousands of feet in length and depth and often forty or fifty feet in thickness. Rain water seeping slowly down through the ground for long ages had dissolved out or leached the copper from the upper portion of these veins, and deposited it again deeper down, making a wonderfully high-grade "enriched" zone that extended from two or three hundred feet to over a thousand feet below the present surface. The Butte veins effectively disposed of any fear of a copper shortage for ten or fifteen years. Yet their development was slow enough so that except for three years of oversupply, copper averaged fifteen cents a pound until 1891.

Luckily the great enriched veins of Butte were in a class all by themselves. If another Butte had been found, the world could not have absorbed its copper for many years. Prospectors hunted for one all through the granite areas of the western United States and the rest of the world. They met with no success. There is only one "Richest Hill on Earth." The high copper price lasted long enough to build the railroad into Butte and to pay for the equipment of the mines for large-scale production. Then they could make fair profits regardless of how low the copper price fell.

In a few years the rapid development of electricity called for even more copper than Butte and the Lake Superior District could produce. Once more industry called for new copper mines. And once more Fate supplied them right on time, with an entirely new style of ore bodies – the high-grade deposits of the southwestern desert.

ENRICHED ORE LENSES

These southwestern deposits were great lenses of ore, usually in limestone. Individual lenses sometimes contained hundreds of thousands of tons, and dozens of these ore bodies were grouped together in areas or districts two or three miles square. Due to the combination of scanty, intermittent rainfall with very deep cañons, surface water has circulated through the porous rocks in the desert areas to a greater depth than is usual elsewhere. In the course of the ages, the circulating water has oxidized and dissolved out or leached the copper from the upper parts of the lenses. Coming in contact with remaining sulphides at greater depth, the copper in solution was precipitated again to form ore several times as rich as the original ore. The resulting bodies of 5 to 20 per cent ore made the greatest high-grade copper districts, save only for Butte, that have been found in the history of mining.

Four major districts of this type were found at almost the same time, between 1870 and 1880. They came to full production just when their copper was needed toward the end of the nineteenth century. Clifton, Globe, Bisbee and Jerome, all in Arizona, were destined to produce as much copper per year as the whole country used at the time they were discovered. The apparent reason why they were developed at so nearly the same time was the opening up of Arizona by the Southern Pacific Railroad, following the pacification or

Mule teams hauled the ore in the early days of Morenci, about 1880. Below is the first Clifton smelter in 1884. *(—Above, Morenci Branch, Phelps Dodge Corp.; below, Mr. J. C. Colquhoun of London)*

extermination of the Apaches. A less obvious reason was the fact that the world needed the new copper and Fate supplied the need.

The Clifton-Morenci District was the first of the southwestern copper camps to become important. Spanish and Mexican explorers had reported the presence of copper in the precipitous mountains north of the Gila River early in the nineteenth century. In 1864, Henry Clifton and a group of American prospectors ventured west from Silver City, New Mexico, into the Apache country. Instead of the gold and silver they were looking for, they rediscovered the rich copper carbonate. At that time the district was so remote that they could make no attempt to locate the mines. Still the story of the copper deposits lingered on around Silver City. Six years later a hardy pioneer and prospector named Isaac Stevens, together with Bob Metcalf and a half dozen others, started on another prospecting trip from the silver camp of Pinos Altos. They soon found striking outcrops of beautiful green copper carbonate near the top of the limestone cliffs two thousand feet above the bed of Chase Creek. They located the first claims in 1872 and founded the town of Clifton in the deep cañon where Chase Creek and the San Francisco River came together.

Copper mining in the Clifton district in the 'seventies was a risky business. The nearest railway point was La Junta, Colorado, seven hundred miles east over desert and mountains. The heart of the Apache country was only a few miles away. Indian raiders killed unwary prospectors as late as 1882. In spite of the dangers and hardships the district grew. Money was needed to build a smelter and to carry the payrolls during the months that must elapse before copper could be hauled out and sold. The Leszinsky brothers of Las Cruces, New Mexico, were ready to gamble on the district. They were ambitious frontier merchants and were used to taking chances on new mines in order to help develop the country. They bought the best claims from Stevens and Metcalf for a few hundred dollars. In 1875 they built a little one-ton adobe furnace to smelt the 20 per cent ore from the Longfellow Mine. The only

At Morenci, small steam engines hauled men and materials on a steep, crooked narrow-gauge track up the canyon from a siding on the broad-gauge line from Lordsburg. Inclines such as those below carried ore down to railway tracks in the gulches. *(—Both, Morenci Branch, Phelps Dodge Corp.)*

fuel was charcoal from the forests on the nearby mountains. Water proved to be too expensive a luxury on top of the mountain, and the following year the Leszinsky brothers built a larger blast furnace plant at Clifton.

Gradually the production from the new district increased. Thanks to a twenty-cent metal price, the Leszinskys made a little money hauling copper out to La Junta. By 1874 the mines looked so promising that they built the first railroad in Arizona, a seven-mile twenty-inch gauge line between Clifton and the mines at the new settlement of Metcalf. Mules were the motive power until 1879, when little wood-burning locomotives were hauled in. Precarious inclined tracks down the steep mountain sides carried the ore from the mines to the railroad. Wire ropes the length of the incline ran from one car up over a pulley on top of the slope and down to a second car. The weight of the descending car pulled the empty one up the two or three thousand feet of 40 or 50 per cent grade. If the rope broke, both the cars plunged down the mountain side and were flattened against the rocks in the bottom. It was a hair raising experience to ride for the first time on a Clifton incline.

In 1875 Captain E. D. Ward, a former army officer and Indian fighter, bought some claims on the slope of Copper Mountain, west of the Longfellow. He interested friends in Detroit and formed the Detroit Copper Company. They built a smelter at Clifton and later moved it to Copper Mountain, seven miles away. There they started the new town of Morenci. By 1880, both the Detroit Company and the Leszinskys were smelting sixty or eighty tons a day of 20 per cent ore, and making fair profits in spite of the long haul to the railway.

The Southern Pacific at last reached Lordsburg, only seventy miles from Clifton, in 1881. Operating the new copper mines became comparatively simple after that. Coke was brought by rail from northern New Mexico to Lordsburg and freighted by mule team to the smelters. On the way out the teams hauled the copper pigs to the railroad. Clifton and Morenci and Metcalf grew into wild border towns, with the

James Colquhoun made a great success of the Arizona Copper Company. *(—J. C. Colquhoun)*

usual faro and poker tables in the saloons, and dance hall girls furnishing the only feminine atmosphere. Still the Clifton district was peaceful compared with Tombstone. New Mexico was so near that many of the workmen were Mexicans. The Mexicans got drunk now and then and shot or knifed one another, but they could never make a town quite so lawless and disreputable as did the pioneer white miners.

The Leszinskys realized that the peak of the copper business had come in 1880. The price of copper was still twenty cents. But the great Butte mines were rapidly coming to production, and the railroad would soon make possible a much larger output from the Arizona camps. The price of copper was bound to drop. Just in time they sold all their interests to a Scotch company. Two million dollars rewarded their struggle against the desert and mountains. The new owners formed the Arizona Copper Company. This company, headed by a fine Scotch engineer named James Colquhoun, became one of the most progressive producers in the Southwest.

Colquhoun started by building a railway to connect with the Southern Pacific at Lordsburg. The cheaper freight made it possible to make a profit out of a great tonnage of lean ore that had been worthless in the bonanza days. This involved a new smelter, larger and more efficient than the old copper-jacketed furnaces the Leszinskys had built. As the rich ore in limestone approached exhaustion, Colquhoun started to prospect the areas of brilliantly iron- and copper-stained porphyry between the older mines. He was rewarded by lenses and broad veins of 4 or 5 per cent ore, far greater than the deposits that had been mined in earlier years, even if they were leaner. As this ore could not be smelted directly, like the limestone ore, Colquhoun devised the plan of first putting the finely crushed ore over the shaking tables that had been used to catch the "sulfurets" in the Colorado gold ores, and then smelting the resulting concentrates. In doing this he was twenty years ahead of his time, using almost exactly the process through which Jackling later made the porphyry coppers so valuable.

This ore train is at Morenci, about 1890, with Chase Creek in the background. Below are a mine office and pre-1900 plant at Morenci.

The photograph of the main shaft on Copper Mountain, Morenci, was taken before 1900. The tunnel, ore bins and mule-drawn train below date from about 1890. *(—Four photographs, Morenci Branch, Phelps Dodge Corp.)*

Bessemer converters were used at the Morenci smelter before 1900. A view of the town and the smelter in March, 1912, are shown below. *(–Above, Morenci Branch, Phelps Dodge Corp.)*

Above are the new Clifton-Morenci smelter and the San Francisco River; below, the Morenci hotel and town, 1912. *(—Below, Morenci Branch, Phelps Dodge Corp.)*

As some of the leaner ore was oxidized and could not be concentrated, Colquhoun worked out a process for dissolving the copper in sulphuric acid, and precipitating it on iron. This first leaching plant, built in 1892, was twenty-five years ahead of the similar process adopted by Chile Copper and New Cornelia. Only a shortage of suitable ore made it a small success instead of a great one.

The progressive policy inaugurated by James Colquhoun was rewarded by more than twenty million dollars in dividends. The fine history of the Arizona Copper Company continued until after the war. Then, faced by the expenditure of several million dollars to allow the treatment of still leaner ore, the Scotchmen sold out to the Phelps Dodge Corporation for a large stock interest.

The Phelps Dodge Company bought Captain Ward's Detroit Copper Company in 1895 and carried on a friendly rivalry with the Arizona Copper Company until the two were combined. Clifton became for many years the greatest copper district in the Southwest. It was one of the most important factors in furnishing cheap copper for the growth of the electrical age.

A hundred miles over the rugged lava mountains west of Clifton Ben Regan found some outcrops of rich silver ore in 1874. He located a claim and called it the Globe. Other prospectors followed and gave their settlement the name of Regan's first claim. For four or five years they made a precarious living trying to mine silver ore in the occasional breathing spells between fights with the Apaches. The silver played out a few feet below the surface. But in 1878 two leasers named Garrish and Van Arsdale found copper carbonate in their open cut on the Old Dominion Vein. The miners were so disappointed at the behavior of their silver ore that for three years they paid no attention to the copper. Even after the completion of the Southern Pacific in 1882, Globe was a hundred and thirty miles by rough wagon road from the railway. This did not encourage copper mining.

In 1881 the Globe Mining Company bought Regan's claims for a few thousand dollars, and built a water-jacketed blast furnace to smelt the copper ore. The Old Dominion Company started at about the same time. The two companies were consolidated in 1882. In spite of the 15 to 20 per cent copper ore, costs were high. The selling price of the copper was low. The Old Dominion Copper Company went bankrupt and was reorganized several times.

A Swiss engineer named Alex Trippel took charge and managed to make a little money in spite of the cost of hauling. Teams of ten or twelve mules took several weeks to make the 260-mile round trip to Bowie on the railroad. The teamsters received $36.00 per ton for hauling coke in, and $15.00 per ton for hauling copper out. In spite of these staggering charges, Trippel made copper for eight or nine cents a pound.

Globe was a small factor in the copper business until Lewisohn Brothers of New York bought control of the Old Dominion in 1895. At that time the Lewisohns were metal dealers, and they wanted their own mines. Without railway connections, no great increase in production from Globe was possible. Therefore the Lewisohns built a branch line to Bowie in 1898. A new smelter and thousands of feet of underground development added to the millions they invested before they were ready to make copper on a big scale.

For the next twenty years, Globe was one of the great copper camps. The new smelter turned out several million pounds of copper a month from the rich carbonate and copper glance ores of the Old Dominion. A river of water was encountered in the deeper workings. Enormous underground pumps managed to keep the mines open most of the time, but now and then a sudden flow would drown the pumps and cause a loss of many thousand dollars. In spite of the great production, the profit from the Old Dominion was comparatively small.

Within a few years the mines had reached a depth of fifteen hundred feet. The ore became lower grade, and profits shrunk still further. The Lewisohns became disgusted with their venture. Dr. Douglas of the Phelps Dodge Company was more hopeful and bought control at a bargain price. Dr. L. D.

Cleopatra Hill at Jerome saw some mining activity, above. The photograph below of the United Verde smelter and mine is from 1885. *(—Both, United Verde Branch, Phelps Dodge Corp.)*

Ricketts took charge for Phelps Dodge. By installing modern machinery he made greater profit from 5 per cent ore than the early operators had made from ore three times as rich. While Globe was not as great a district as Clifton or Bisbee, it added hundreds of millions of pounds of copper to the world's supply.

Bisbee was discovered a few years later than Globe. Dr. Douglas' engineering skill, combined with the luck that is necessary for any mining success, brought the Copper Queen to full production in half the time that was required in the other Arizona districts. By 1890, Bisbee was one of the world's great sources of copper. A few years later, when Cap'n Jim Hoatson and his friends found the rich ore bed under the limestone capping in the Irish Mag, the production shot ahead until this district alone was producing twelve million pounds of copper a month — as much as all the mines in the United States yielded in 1880. Bisbee was by far the greatest of all the southwestern high-grade districts, the perfect example of the enriched lenses that characterize the ore bodies of this style.

Finally came Jerome and the United Verde. The first one to notice the big outcrop of copper-stained iron oxide high up on the mountainside across the brilliantly painted Verde Cañon from the San Francisco Peaks was Al Sieber. He was one of the army scouts who finally brought success to the long campaign against the Apaches. Sieber stopped long enough in 1877 to stake a claim that he called the Verde, because of the green carbonate stain. George Hull came the same year and located adjoining ground. A few more prospectors gradually drifted in. By 1880 the district had acquired fame enough so that the Phelps Dodge Company sent Dr. Douglas to examine it. The 175-mile wagon haul across mesas and deeply carved cañons to the Santa Fe Railroad discouraged Douglas. He reported that there was a little copper ore, but he could not see that it was worth anything.

This was a hard blow to the struggling camp. Still the prospectors kept on digging. Under the black pillars of iron

A Marion steam shovel loads cars at the beginning of work on the United Verde open cut. Trucks were employed at a slightly later stage of development. *(—Both, United Verde Branch, Phelps Dodge Corp.)*

oxide on the Wade Hampton claim the McKinnon brothers and M. A. Ruffner found copper glance that assayed 30 per cent. Governor Tritle of Arizona bought them out for a few thousand dollars. The job soon proved too expensive for Tritle. After much difficulty he persuaded Fred Thomas, a San Francisco engineer and promoter, that the rich copper ore might someday become valuable. In the summer of 1882, Thomas paid $500 cash for an option on the Wade Hampton and adjoining claims. The rest of the $15,000 purchase price was due on December 1st. This was much more than Thomas could raise. George Treadwell of San Francisco, who later developed the great Treadwell gold mine in Alaska, finally came to the rescue. Thomas succeeded in having his payment reduced to $7,500, which Treadwell advanced. The money came just in time to save the option.

The next year Thomas organized the United Verde Copper Company. Its secretary was Eugene Jerome, for whom the growing town was named. Thomas managed to scrape together enough money to build a fifty-ton furnace. With this he turned out nearly $800,000 worth of copper in the first year, and paid $62,000 in dividends. Then the price of copper dropped, and the mine had to shut down. Even 20 or 30 per cent ore was of no value in such a remote camp when copper sold for less than ten cents a pound. It looked as though Dr. Douglas had been right.

There was one man who was willing to take a chance in spite of the hard times in the copper business. In 1888, W. A. Clark came down from Butte with his smelter man, Joe Giroux. One look at the rich ore on the Wade Hampton was enough. Clark took a lease on the United Verde and bought it the following year. The development work he carried on soon proved that the 10 to 20 per cent copper glance ore was two hundred feet wide and six or eight hundred feet long. Even in Butte this would have been a great ore body.

In 1894 Clark built a forty-mile narrow-gauge railway to connect with the new Santa Fe line running south from Ash Fork toward Prescott. The little road with the ambitious name

In its final stage the United Verde open cut was operated by Phelps Dodge. Below is the United Verde Extension smelter in 1923. *(—Above, United Verde Branch, Phelps Dodge Corp.; below, Arizona Historical Society Library)*

of United Verde and Pacific wound around the steep mountainsides like a tortured snake. But it hauled ore and coke a lot more cheaply than the slow mule teams. With a new smelter and roast heaps like those at Rio Tinto to burn the sulphur out of the ore, the United Verde was soon one of the great copper mines. Its production even rivaled that of Bisbee. As the grade of the ore dropped with increasing depth of the mine, new equipment and larger tonnage kept up the yield of copper and the profits. In the thirty years before he died, Senator Clark made $60,000,000 out of the mine that Thomas had bought with so much difficulty for $8,000.

To offset his wonderful success with the United Verde, Destiny played one of her most amusing tricks on Senator Clark and gave the cream of the ore body to an insolent intruder. Just below the "Big Hole" is a great fault. The rock east of this Jerome Fault has slid down toward the Verde cañon for half a mile. As a result, the limestone and lava that were laid down on an ancient erosion surface long after the ore body was formed are high up on top of the mountain west of the fault and of the United Verde, and far down toward the cañon east of them. Under these comparatively recent rocks east of the fault, the older schist that contains the ore is buried six hundred feet deep. The fault cut off the greatly enriched top of the United Verde ore body, together with the enclosing rock, and slid it two thousand feet down and an equal amount to the east. Erosion then exposed the roots of the ore west of the fault. The former top of the ore body, now east of the fault, remained safely covered and hidden by the limestone and lava.

Small chunks of ore were dragged down along the plane of the fault, and were found on it at the surface in United Verde ground. The dip of the fault would soon carry it east, under claims owned by George Hull, a Jerome pioneer. Hull thought the Jerome Fault was a big new vein. In 1899 he formed the United Verde Extension Mining Company and induced a New York broker named Louis Whicher to sell a lot of stock and to sink a shaft through the lava on the Little

Daisy claim. Neither of them recognized that if the fault had been a vein, it would have belonged to Clark because of the apex law.

In the next twelve years Whicher and his various associates raised and spent nearly half a million dollars on the Verde Extension. Only a few streaks of ore resulted from the development. The Little Daisy was finally shut down. Just because it had failed it acquired the general reputation of being a rather odoriferous stock swindle.

Among Verde Extension's stockholders was Major Andrew Jackson Pickrell, a fine old southern gentleman from Prescott, Arizona. The Major still had faith that somewhere in the Verde Extension there was a big ore body. He wrote to his friend James S. Douglas about it. Jim Douglas – a son of Dr. James Douglas of the Copper Queen – had just finished earning the soubriquet of "Rawhide Jimmy" by his relentless economy as manager at Nacozari and later at Cananea. He was looking for a chance to get a mine of his own, to live up to his distinguished father's reputation.

Major Pickrell's story sounded good to Jim Douglas. With his friend George Tener of Pittsburgh he sent an engineer to examine the Verde Extension. The engineer reported that there was nothing in the vein theory, but there was a fair chance of finding the very valuable faulted top of the United Verde. It was a good gamble – but they must stay away from the fault. Otherwise they might be in for a dangerous apex suit.

Douglas and Tener started development. Two hundred thousand dollars went into the ground, with no results. They decided to risk another hundred thousand. Two years after they started work, a crosscut on the twelve hundred level found five feet of 45 per cent copper glance. That would pay expenses anyhow – and they started to sink again.

In 1916, after a four-year campaign of development, the fourteen hundred level electrified the mining world by cutting three hundred feet of 15 per cent ore. It was the faulted top of the great United Verde ore body and its richest spot. A vertical sideline agreement with the old company prevented a complicated apex suit that might have ruined the Verde Extension. Dividends began the year the big ore body was found. Under Clark's very nose, Jim Douglas and his friends have taken out $42,000,000 in dividends.

After adding these four great high-grade districts to the list of copper producers, Fate rested a bit. It was high time. Between Butte and the southwestern deposits, United States copper production increased until the 1897 output was eight times that of 1880. The great copper companies were sure they were ruined. No demand could possibly keep up with such a flood of copper. The price of the metal dropped steadily from twenty-two cents in 1880 to eleven cents in 1885. Save for a short breathing spell late in the 'eighties, it stayed below twelve cents until 1898. Woe, woe, cried the copper magnates. All their efforts in finding the copper the world so craved were doomed to go unrewarded. A handful of the richest mines might still pay dividends, but the rest of them were sunk into hopeless ruin.

As usual, the copper kings could not see beyond their noses. The low price of the 'eighties and 'nineties was the best thing that could possibly have happened to the copper miners. It made possible the swift and universal adoption of electricity. Due to it, the copper business grew from the insignificant trade of a few thousand obscure miners and metal workers to one of the greatest industries in the world.

Late in the 'nineties the reward came. Consumption increased at an incredible rate. The price of copper jumped to seventeen cents. Instead of a hopeless glut, the copper kings began to cry famine. Dividends went sky-high, and stock market prices still higher. The rise of "Amalgamated" focused the attention of promoters as well as miners on the new giant among metals. An army of engineers and prospectors and stock gamblers set out to find new copper mines.

Naturally they looked for mines like the ones that were enjoying the spectacular success. The high-grade "enriched" desert ore bodies were still in style. Soon Arizona and New

Mexico and northern Sonora were alive with copper hunters. It was a poor prospect that had not been examined half a dozen times. Many prospectors became skilled raconteurs, exposing the foibles and eccentricities of a regular directory of famous engineers who had visited them — and turned them down.

It was not the engineers who won the honors in this recurrence of the fashion of desert copper mines. Maybe they were looking for ore bodies just like those they had seen, forgetting that nature never exactly repeats herself. Or maybe it was just the perversity of chance. Anyhow the big prizes in this ore hunt went to the gambler and the practical miner. Bill Greene found Cananea and Cap'n Jim Houston found the Irish Mag. All the geologists could do was to tell why the ore was there after it had been found.

Dr. Ricketts did vindicate the engineering profession to a certain extent. The Doctor was running the Old Dominion Mine in Globe and looking at prospects for Dr. Douglas and the Phelps Dodge Company on the side. On one of his examining trips south of the Mexican border in 1904, he happened to spend the night at the old camp of Nacozari, in Sonora. The Spanish Padres had mined silver at Nacozari nearly three hundred years before. When the hunt for copper was at its height the Guggenheims took over the old mines and started to work them for copper. Curious to see how a silver-lead smelting company would treat a copper property, Dr. Ricketts took a day off to visit the Nacozari mine. High on a bare ridge, far above the little town, he found great rugged outcrops of brilliantly iron- and copper-stained volcanic breccia — "The Pillars of Nacozari," the Mexicans called them. In a tunnel three hundred feet below the outcrops, the Guggenheims were mining beautiful enriched copper glance ore. The mining end of it looked good. But the treatment of the ore spoiled it all. They were trying to concentrate this copper ore in a little mill as though it had been high-grade silver ore, and were hauling the concentrates eighty miles by mule team to the railroad and shipping them to El Paso. No copper mine could stand that sort of treatment. Still it was none of his business, so Dr. Ricketts thanked the Guggenheim engineer for his courtesy, and went away and said nothing.

It did not take the Guggenheims long to decide that they were not ready to go into the copper business. One of the partners called on Dr. Douglas in New York. He said that a Phelps Dodge engineer had visited Nacozari, and seemed to like it. Would Dr. Douglas care to buy the mine?

Dr. Ricketts advised the purchase. So the Phelps Dodge Company paid the Guggenheims the amount they had lost on the venture. A campaign of development proved that many million tons of 3 per cent ore underlay the Pillars of Nacozari. Dr. Ricketts built a railroad to Douglas, Arizona, and designed a six hundred-ton concentrating mill like that at Clifton. In 1906 production started. Nacozari became the fourth big mine of the Phelps Dodge Corporation.

One more great lens of copper glance was found before the high-grade ore bodies went out of style. It was far from the desert in every way. Chance was responsible for its discovery, as no one would have been foolish enough to hunt for copper in such a place. Gold was the metal that fired the imagination in Alaska in 1900. Yet copper had been known for a long time in Alaska. Boulders of rich copper ore had been found years before, carried far from their home by glaciers. Because of them the swift river that emerges from the rugged mountains a few miles south of Cordova was called the Copper River. The Indians had named a tributary of this river the Chitina, which means "copper stream." In 1899 a venturesome party of prospectors headed by Warner and Jack Smith followed the river toward its source, hunting for placer gold. The search was unsuccessful. But high up on the crest of Bonanza Mountain they found a mass of almost pure, steely copper glance twenty to forty feet wide and four hundred feet long. This fantastically rich ore assayed over 50 per cent copper. In the steep talus slope below the outcrop were thousands of tons of fragments of ore, torn off by the ice all ready to ship.

The prospectors located a few claims and returned to the coast to tell of their discovery. Their friends just laughed at them. What good was copper at sixty-five hundred feet altitude at the foot of Mount Wrangell, in the very home of blizzards and perpetual ice? To get it, a railroad would have to cross the mossy swamps of the broad river delta, pass for miles over the solid ice of the great coastal glacier, and then carve its way through the rocky cliffs of the upper cañon. No copper mine could be rich enough to stand that expense. It was doubtful if the railroad could be built at all, and without it there was no possible way to get the copper to the coast.

For seven years the Bonanza ore body was allowed to stand untouched save for shallow tunnels. The prospectors who had located it had to take in partners to help carry on the annual assessment work. Few sets of partners lasted more than one year. Before long the title was involved in lawsuits so complicated that a solution seemed hopeless.

It was only in 1907, when another "copper famine" scare raised its head, that any company ventured to attack the almost impossible problem. Then the Guggenheims decided to risk on the Bonanza some of the millions they had won in Mexican lead and silver mines. They bought the property, and formed the Kennecott Copper Company. J. P. Morgan was persuaded to help finance the railroad. It took three years to build the 195-mile line — three years and $13,000,000. Crossing the glacier, the rails were slowly and inexorably swept down by the ice and had to be relaid every few weeks. It was the most daring and the most expensive railroad ever built. A rope tramway nearly three miles long, with a four thousand foot drop, brought the ore down from the mine to the railroad. A mill to treat the leaner marginal ore and the houses for workmen were placed in niches carved in the rock of the ridges, where the deadly snow slides could not reach them. There seemed no end to the millions that must be spent.

As soon as the railroad was finished, the millions came rushing back again. For a time Kennecott became the bugaboo of the copper producers. As production kept mounting, the rival companies feared that the Bonanza ore would wreck the copper market of the world. New ore bodies almost as rich as the first one added to the fear. Then Fate relented. Ore reserves dwindled and the grade of ore fell. While the original Kennecott ore bodies have returned far more than the colossal investment, in the end they have proved to be of little importance compared with the less spectacular low-grade mines that now share the name of Kennecott.

With the Bonanza, the list of the great enriched copper lenses is complete. Tens of millions of dollars were spent hunting for more of them. At Silver Bell, Helvetia, Courtland, and a dozen other Arizona camps shallow bodies of rich ore encouraged the hope that another Bisbee had been found at last. But the ore bodies turned out to be small, and two dollars were spent for every dollar taken out of them.

Courtland looked so good that the Phelps Dodge Company and the Calumet and Arizona almost fought for claims, to the great delight of Bill Holmes and the other prospectors who received high prices for their locations. W. J. Young, a lumber man from Clinton, Iowa, got the best of the camp against his will in the Mary Mine, which he took over in payment of a bad debt. The Calumet and Arizona wasted a few hundred thousand dollars on Bill Holmes' Leadville claims and then found a continuation of the Mary ore in the Germania. The upper levels of the Mary and the Germania had as sweet a high-grade ore body as an engineer ever dreamed of. The Southern Pacific and the El Paso and Southwestern rushed in to grade branch lines, each trying to block the other from entering the wonderful new district. Three townsites sprang up to house the thousands who joined the rush. Men stood in line all night to buy lots in a new subdivision. The saloons and dance halls were like the good old days in Tombstone and Bisbee.

One brief flash of greatness, and it was all over. For Fate had played an unkind trick on Courtland. She had picked up a block of rocks half a mile square, including all the rich ore bodies, and had shoved it five or six miles southeast of

Daniel C. Jackling proved the worth of the disseminated deposits at Bingham, Utah. *(–Utah Copper Division, Kennecott Copper Corp.)*

its original position. In the place where the block had originally been, the roots of the ore bodies were obliterated by a mass of molten rock that later solidified into a porphyry containing no trace of copper. As soon as the Mary and the Germania sank three hundred feet they got below the wandering block of rock in which the ore occurred and encountered the hardest and barrenest limestone that ever dulled a drill. The ore bodies lasted only a few brief years. Young's Great Western Mining Company made a profit of a few hundred thousand dollars out of the Mary and lost it all hunting for more ore. The Calumet and Arizona got back most of the money it spent in the Germania. The rest of the investments in Courtland were a complete loss. There are just two stranded families left in Courtland now, in the midst of the crumbling adobe walls and flapping strips of corrugated iron that are the mausoleums of ten thousand hopes.

Like hoop skirts and pantalettes, the enriched desert ore lenses had gone out of style. No amount of hunting could find another of them.

DISSEMINATED OR PORPHYRY ORE BODIES

By 1907 everyone was talking copper famine again. Consumption had passed all that Butte and Bisbee and other high-grade camps could supply, and there would be no more Buttes and no more Bisbees. All the experts solemnly declared that copper could never sell below twenty-five cents a pound as long as man used electricity.

Then along came Daniel C. Jackling and upset all their calculations with an entirely new fashion plate for copper ore bodies to follow. For once, chance had very little to do with it. One simple little idea that came out of Dan Jackling's head doubled the world's supply of copper. With that idea he made the disseminated, or porphyry ore bodies and made himself the leader of all the engineers of copper.

Jackling's bitterest enemy really started it. Colonel – by courtesy – Enos A. Wall came into Bingham as a prospector and trader in 1887. Bingham, Utah, was an old camp even

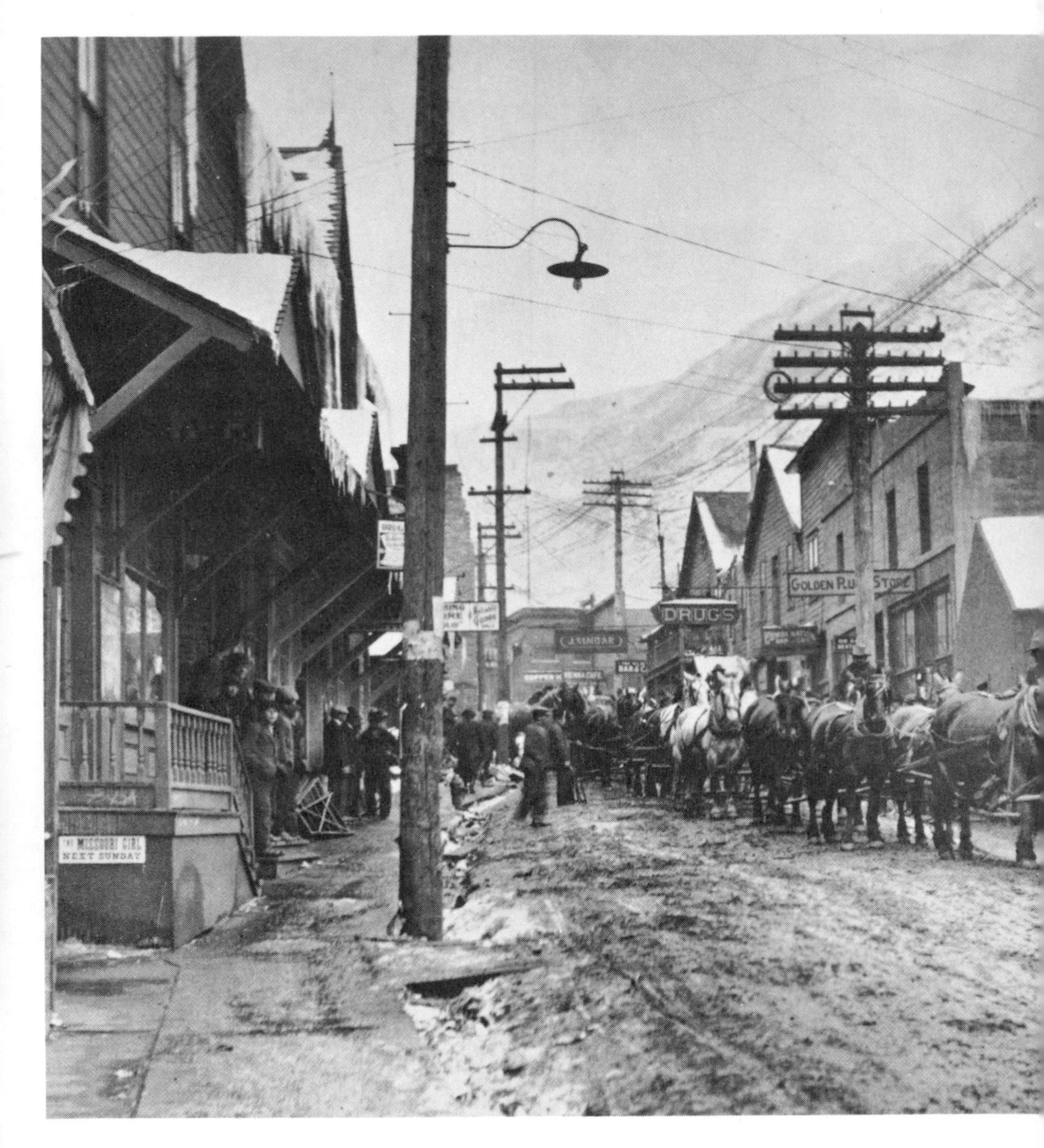

Main Street in Bingham was still mud in 1915, long after Utah Copper was a success. *(–Utah Copper Division, Kennecott Copper Corp.)*

then. Way back in 1863 General Connor had taken a party of army officers and their wives on a picnic in a steep cañon in the Oquirrh Mountains east of the Great Salt Lake. One of the ladies picked up a glistening rock that was heavy as lead. No wonder, for it was almost pure lead carbonate carrying rich silver values. General Connor established the rules of the new Bingham Cañon mining district, and ever since then lead and silver and gold mining have flourished in the limestone beds that form a half-circle around the center of the cañon.

Within this half-circle the precipitous talus slopes were underlain by bright yellow and red iron-stained porphyry that the early miners thought had no value. Colonel Wall had a different idea about it. The gravel in the bottom of the cañon was stained green with copper carbonate. In the porphyry itself there was copper stain, and one outcrop three hundred feet wide assayed 3 per cent. Wall had been in Butte and thought the Bingham showing beat anything in the "Richest Hill on Earth." He relocated the claims that the earlier prospectors had abandoned, and from that day on every dollar he could raise went into buying more ground and running more tunnels. By 1895 he had secured two hundred acres and had spent $20,000.

This copper ore that Colonel Wall developed was entirely different from any that had been known before. It was not in beds of lava or conglomerate as in the Lake Superior District, or in sheets or veins as in Butte. It was not in rich lenses like the southwestern enriched deposits, or in big masses of solid sulphide like Rio Tinto. The Bingham ore was a whole mountain shot full of specks and crisscross streaks of copper sulphide. The copper near the surface had been dissolved out by surface water, leaving a barren oxidized "capping" a hundred feet thick. For another hundred feet below the capping, the copper was deposited again by the descending water that had leached out the surface. The resulting zone of "enriched ore" was unusually high grade — often assaying 3 per cent copper. It faded away gradually into the unenriched or "primary" ore that ran 1 to 2 per cent copper, and this in turn

R. C. Gemmell was associated with Jackling in starting to mine Bingham's disseminated ore. *(—Utah Copper Division, Kennecott Copper Corp.)*

faded away into barren rock. There were no walls, and the limits of mining were set by the gradually decreasing grade of ore. The whole porphyry intrusion that had been forced up from the depths of the earth while still molten was mineralized.

This disseminated ore could not be smelted like richer copper ores, as it would cost a prohibitive amount to melt the 97 per cent of rock. But Colquhoun at the Arizona Copper Company had learned how to separate copper sulphides from the finely ground rock on concentrating tables. Wall thought the same process would work on his ore, and he built a little mill to prove it.

All of this took money. Wall had spent all he had and now he was ready to get some help. So he wrote to Captain J. R. De Lamar – a Dutch adventurer who had made himself the biggest lead and silver mine owner in southern Utah. De Lamar sent his manager Cohen to examine Wall's prospect. A stamp mill and concentrating table test gave a 30 per cent concentrate that could be easily smelted. But the price of copper had dropped to less than ten cents and Wall wanted $375,000 for his property. De Lamar gave up his option.

Cohen was not satisfied. Three years later the price of copper had risen to twelve cents a pound. Cohen induced Wall to give another option, and De Lamar sent a young engineer named Robert Gemmell and a young millman named Daniel C. Jackling to sample and test the ore. The price still looked too high, and De Lamar dropped the option again.

Jackling and Gemmell did not agree with the decision. A few months later, when Victor Clement replaced Cohen as manager for De Lamar, they persuaded Clement to take still another option. Jackling estimated that there were 12,000,000 tons of 2 per cent ore on the property, and he could save 70 per cent of the copper in a 22 per cent concentrate. De Lamar made a $50,000 payment for the first quarter interest. Then he had a fight with Clement and fired him, and threw up his option rather than pay Wall $250,000 for another quarter interest.

Cohen came back as manager for De Lamar. He wanted to realize on the quarter interest that Clement had bought. He successively offered the Wall property to Ben Guggenheim, to Charles A. Coffin of the General Electric Company, to John Hays Hammond, and to W. A. Clark of Butte. They all turned it down. No one could make any money out of 2 per cent ore way out in a cañon in Utah.

Jackling left De Lamar and in 1902 went to Colorado to work for Charles MacNeill in the mills that treated the Cripple Creek gold ore. Jackling was then thirty-two years old — not very experienced, but full of energy and determination. He told MacNeill and Spencer Penrose, the leading mine-owners in Colorado, that he could make them a lot of money out of the Bingham low-grade ore. They sent F. H. Minard and R. A. F. Penrose to make a check examination and were converted. Jackling bought the De Lamar quarter interest for them for $125,000, and optioned another 55 per cent interest from Colonel Wall for $400,000. In June, 1903, they formed the Utah Copper Company and set Jackling to work to build a mill with three hundred tons a day capacity.

The distinguished engineers on swivel chairs in luxurious New York offices chuckled to see MacNeill and Penrose make such fools of themselves. But it was rather too bad to see a promising young engineer like Jackling ruin his career with such a mistake before it was well started. The Arizona Copper Company had given the crazy new concentrating process a thorough trial and could hardly make it pay with 3 per cent ore in an established camp, with a railroad all built. What chance of success could Jackling have with 2 per cent ore in a camp where he must either build an expensive fifteen-mile mountain railway or haul his concentrates out by mule team?

The *Engineering and Mining Journal,* the infallible leader of technical papers, buried the new enterprise in ridicule with an editorial that said: "It would be impossible to mine and treat ores carrying 3 per cent or less of copper at a profit under existing conditions in Utah. On the Company's own showing, therefore, the more ore it has of the kind it claims, the poorer it is."

Jackling refused to stay buried. He had an idea that made obsolete all the theories of all the swivel-chair graybeards in the business. They were quite right in thinking that his three-hundred-ton mill was going to lose money. But he knew – and they did not know – that if he increased the size of his mill to five thousand tons a day, he could turn the failure into a great success.

This was Jackling's contribution to the world. He saw that most of his cost – railroad, mine development, power lines, administrative expense, town, shops and so on – was independent of the amount of ore treated. If he could divide this cost by five thousand instead of by three hundred, it would become insignificant. This simple bit of arithmetic gave the world a hundred billion pounds of cheap copper – more than it has used in all the ages since the first cave man picked up the first copper pebble. This one idea brought Jackling the highest honors the engineering profession can give.

MacNeill and Penrose were ready to back Jackling to the limit. They persuaded the Guggenheim Exploration Company and Hayden, Stone and Company to join them in financing this expensive venture. The public helped, by buying Utah Copper stock at ten dollars a share. Jackling spent eight million dollars before the big mill was ready to run in 1907. He paid three millions more for the adjoining Boston Consolidated property, that proved to have a continuation of the ore body. If Jackling failed, he would at least have the biggest failure in mining history.

Many engineers were still skeptical. Colonel Wall himself went over to the opposition. Until 1908, he had retained his interest and was a director. But it made his blood boil to see an upstart like Jackling get all the recognition. Wall had found the ore body, and he wanted to run the company in his own way, without recklessly spending so many million dollars. At last Wall resigned as director and devoted all his energy to fighting Jackling and Utah Copper. He even established a well-edited mining journal – *Mines and Minerals* – that proved

From left to right, leaders of the Utah Copper group, about 1910, were Sherwood Aldrich, Spencer Penrose, Charles M. MacNeill, D. C. Jackling, Charles Hayden, A. Chester Beatty and Allen H. Rogers. Below is a view of Kennecott's open pit at Bingham about 1965; benches are generally 75 feet high. *(—Both, Utah Copper Division, Kennecott Copper Corp.)*

mathematically for three years and hundreds of pages that the new "porphyry coppers" could never pay back the gigantic investment.

The criticism was drowned in a flood of dividends. Jackling's estimates were more than realized in the first year. He soon increased the profits still more by stripping the barren capping off the ore body with steam shovels, and then mining the ore itself in a colossal open cut, by five-ton mouthfuls. Bingham became the most spectacular of all the works of man. The shovels were eating away a whole mountainside of bright green and gray and red rock, two and a half miles long and a half mile high. From across the cañon the trains and the hundred-foot shovels on the forty terraces looked like little toys, so great was the scale of the operation.

The five-thousand-ton mill that seemed so outrageously big was doubled, and doubled again three times over, in geometrical progression. In one day the Utah Copper shovels mined seventy thousand tons of ore, and the mill turned out more than a million pounds of copper in the high-grade concentrate.

Utah Copper installed all the newest machines for mining and treating the ore as soon as they were invented, and many of them were invented by Jackling and his assistants. The mill recovery that started at 60 per cent increased with the advent of flotation process to 90 per cent. The cost declined so that even when the grade of ore dropped to 1 per cent copper, Utah Copper was turning out the metal for five cents a pound.

In the first twenty-five years after Jackling started his first porphyry copper mill, Utah Copper produced four billion pounds of copper and paid over $200,000,000 in dividends. The original stockholders received $120 per share on their $10 investment. And with 600,000,000 tons of ore developed, they could be sure that once the Depression was over, the lowest-cost copper mine in America would continue to pay dividends for forty or fifty years.

The copper hunters were quick to take advantage of Jackling's new style of ore body. Before Utah had paid its first dividend, the mining countries of the whole world were alive

W. N. McGill was one of those who started the Nevada Consolidated, later acquired by Kennecott. Below, a steam shovel works in the first cut at the Eureka Pit, Copper Flat, near Ruth, Nevada, on April 5, 1908. This was later merged into the larger Liberty Pit. *(—Above, University of Nevada, Reno, Special Collections Library; below, Nevada Mines Division, Kennecott Copper Corp.)*

with engineers hunting for the big red and yellow iron-stained areas that meant disseminated copper deposits. The spectacular surface coloring made the search a rapid one. Within five years after the Utah mill started they had found a dozen great low-grade ore bodies. Then porphyry ore bodies went out of fashion. Hundreds of millions of tons of lower-grade ore surrounding the richer kernels were added to the estimated reserves as the progress of mining and ore treatment reduced the costs. But in the twenty succeeding years only a handful of small and lean deposits of comparatively little value had been added to the list.

Ely, Nevada, was the first district to follow in Utah's footsteps. Like Bingham, it had been a gold and silver camp since the 'sixties. But it had never amounted to much. About 1900, Senator Clark of Butte took an option on a few claims that had copper carbonate outcrops. A little work convinced him that it was not another Butte, and he dropped the option.

A local sheep and cattle rancher named W. N. McGill then turned miner, and in 1901 found a little 3 per cent copper ore. The next year Gray and Bartley formed the New York and Nevada Copper Company to prospect other claims. It found what it considered to be broad veins of 2 to 3 per cent ore.

A violent strike called by the Western Federation of Miners nearly wrecked the district in December of that year. The superintendent of the New York and Nevada, J. A. Traylor, refused to be intimidated. A group of six armed strikers came to his office one morning, and said they were going to take him out and hang him. Traylor shot through the door, and killed four out of the six. That ended Ely's labor troubles.

By this time the success of Jackling's experiments at Bingham was recognized by the more progressive engineers. In 1903, Mark L. Requa bought a group of claims adjoining the New York and Nevada, and incorporated the White Pine Copper Company. Senator Clark's old smelter man, Joe Giroux, found ore in the Pilot Knob and Brooks claims, and turned them over to the Giroux Consolidated Copper Company, destined to see some of the wildest financing in the history of mining. The Butte and Ely and the Cumberland and Ely Copper Companies sold a lot of stock in Boston and New York, and found low-grade ore. A wild boom in Ely copper stocks ensued. The Nevada Consolidated Copper Company bought the New York and Nevada and White Pine properties and started a race with Cumberland and Ely to acquire water rights and mill sites. Stock in these two companies sold for tens of millions of dollars before a mill had been even started.

In 1905, J. Parke Channing examined the Nevada Consolidated, and reported that it contained 26,000,000 tons of 2 per cent ore. On his recommendation the Company built the 140-mile Nevada Northern Railroad up the Steptoe Valley to connect with the Southern Pacific. Nevada Consolidated then absorbed Cumberland Ely, under the leadership of William B. Thompson and James Phillips. The new company planned a 2,500-ton concentrator and increased it to 10,000 tons before construction started in 1906. Stripping the lean capping off the Eureka and Liberty ore bodies by steam-shovels began the same year. Two years later the steam-shovels started to mine the ore, and the next year Nevada Consolidated was operating at full capacity. The success was so great that Utah Copper was glad to buy a half interest in 1910 for the bargain price of $4,500,000 in stock. The Nevada Consolidated mine paid $70,000,000 in dividends in the next twenty years.

Development of the other porphyry coppers rapidly followed. In 1906, Jackling and his associates found the next one at Ray, in a hot, desolate Arizona cañon east of Phoenix. Six years before an English company had developed 200,000 tons of ore that they thought ran 4 per cent, but that really contained only 2 per cent copper. Naturally the company failed, after amusing the desert with half a million pounds sterling worth of wasted work, made colorful by the incongruous tea and tennis costumes of the English engineers. Seeley W. Mudd optioned the Ray mines for the Jackling interests and found 50,000,000 tons of 2 per cent ore. The 10,000-ton concentrator started in 1911, after $10,000,000 had been spent.

Jackling and his friends made another great success with the Chino Copper Company at Santa Rita, New Mexico. Spaniards, Mexicans and American pioneers had mined rich streaks of copper ore in Santa Rita ever since the sixteenth century. In 1900, A. C. Burrage and others of the "Amalgamated crowd" bought the principal property and tried unsuccessfully to develop another Butte. Then the Santa Rita Copper Company was shut down for a few years. John M. Sully had worked for the Amalgamated interests at Santa Rita and thought they were missing a trick. In 1906, he made a careful examination, and came to the conclusion that instead of a high-grade mine, Santa Rita contained tens of millions of tons of 2 per cent ore. It took Sully three years to convince anyone else that he was right. At last Jackling and his Hayden, Stone friends decided to try it. They formed the Chino Copper Company, developed the ore by hundreds of churn drill holes, and built another 10,000-ton mill. In 1911, Chino joined the ranks of the great producing copper mines. John Sully stayed on as Jackling's manager and remained in charge after twenty-six years as benevolent despot of Santa Rita.

Miami and Inspiration came next. The hundreds of acres of red hills a few miles north of Globe had long ago attracted the attention of prospectors because of the streaks of rich copper carbonate ore that occurred in them. The streaks all played out near the surface, and the district looked hopeless. Then, in 1904, a stock company of doubtful reputation called the Inspiration found disseminated chalcocite ore. It tried to make money with a fifty-ton mill and failed. J. Parke Channing heard of the operation and came to Globe for the Lewisohn interests to see if there might be a low-grade mine like that he had examined at Ely. He did not get the Inspiration, but did succeed in persuading a famous Arizona prospector known as Black Jack Newman to give him an option on the adjoining property. There is the usual story that Channing was ready to quit after one shaft found lean material and a second shaft looked no better, but Black Jack put in one more round and broke into the ore. The Miami Copper Company resulted, and

At Inspiration's mine and plant about 1916, left to right are the ore bins, twin shafts and crushing plant, and hoist and compressor house. The Inspiration mill is below, with the International smelter in left background. *(–Both, Inspiration Consolidated Copper Co.)*

in 1911 was milling 3,000 tons of 2.5 per cent ore at a beautiful profit.

The original Inspiration Company gave place in 1908 to another backed by W. B. Thompson, who had acquired fame and wealth by manipulating stock in the Nipissing silver mine and in Cumberland Ely. In the next few years Inspiration found 20,000,000 tons of ore and absorbed the Live Oaks Company, with as much again. The ore was harder to mill than that at Miami. Financing was too much for the Thompson group, who sold out control to Anaconda. Dr. Ricketts became consulting engineer. He did not think the mill Thompson's engineers had planned would recover enough of the copper. To the horror of the stockholders he threw away a million dollars' worth of mill construction, and spent a year and another million dollars experimenting. Then he built the first mill that used the new flotation process. This remarkable method of concentrating metallic minerals from low-grade ores depends on the fact that oil sticks to the surface of the metallic particles more tightly than to the worthless particles of finely pulverized rock. If air bubbles are blown through a pulp or thin mud of powdered ore and water containing a minute amount of oil, the oil picks up the metallic grains and the air bubbles pick up the oil, so that a froth of air and oil and the metal-bearing minerals floats to the top of the tank and flows off as a rich concentrate. This process increased the recovery in copper mills by 50 per cent. The result of Dr. Rickett's delay and introduction of flotation at Inspiration was that this company caught the high copper price of 1915 with the most successful mill that had ever been built, instead of struggling along with a plant that was not suited to the ore.

Dr. Ricketts and his friend John Greenway had another triumph at the New Cornelia property at Ajo, which the Calumet and Arizona developed. With an entirely new leaching process, Ajo was a complete success from the day the plant started in 1917.

The Phelps Dodge Company was not to be outdone by its old rivals. It developed 20,000,000 tons of porphyry ore in

Sacramento Hill, in the middle of the great Bisbee ring of limestone ore bodies. Unfortunately, 20,000,000 tons of ore is hardly enough to stand the enormous expense of equipping a low-grade copper mine. Sacramento Hill has had only an indifferent success. In the Clifton-Morenci district also the Phelps Dodge Company found that hundreds of millions of tons of low-grade ore lay between richer lenses. This Clay ore body at Clifton is, next to Utah, the largest of all the American copper deposits. It will take $40,000,000 to equip it on the scale it requires, and $40,000,000 is more than the ore body is worth when the world has more copper than it needs. The Clay ore body must wait until copper gets scarce again.

The United States had all the luck in finding new copper mines for forty or fifty years. In 1870 this country produced only about an eighth of the copper used in the world. By 1910 it was turning out three-fifths of the enormously increased output. The old mines in Germany, Spain, Japan and South Africa that had supplied the world with copper for hundreds of years increased production when electricity made copper really worth mining, but for more than a generation no really great copper ore body was found outside of the United States. It was no wonder that the American Copper Kings looked on the world as their plaything.

The luck was too good to last. The cycle of American copper discoveries ended in 1913. In the succeeding twenty years, foreign countries have made all the great finds, while the proud American companies were forced to delude themselves and their stockholders by adding to their ore reserves hundreds of millions of tons of lean material that could be valuable only with the high copper price that followed the World War. For a time the owners of American mines kept control of the situation by buying the new mines in other countries. As more and more deposits were found in remote lands, this became impossible. The God of Luck has taken away our copper supremacy forever.

THE SOUTH AMERICAN MINES

Cerro de Pasco started the avalanche of foreign copper. This district high up in the Peruvian Andes was one of the oldest in the new world. Ever since the days of the Incas it had been a great producer of silver. The Spanish and Peruvian owners of the hundreds of little properties on the desolate mountain 14,000 feet above sea level had mined forty million tons of ore from rich stringers near the surface. The whole mountaintop was honeycombed with their drifts and stopes. They packed the ore to the little mud furnaces on the backs of hardy llamas — the camels that turned south from their point of origin in North America and were stunted by the Andean hardships while their luckier cousins crossed into Asia and grew into the camels we know. Other trains of llamas, with their brightly clothed Indian drivers, carried the bullion two hundred miles over precipitous trails to the capital of Lima and to the port of Callao.

A hundred feet below the surface the silver ore gradually turned to copper. At first this new ore was so rich in silver that the Peruvian miners could still work it by their primitive methods. With greater depth the silver content decreased until the old-fashioned mining was no longer profitable. In that remote camp in the heart of the Andes, modern large-scale operations seemed out of the question. Cerro de Pasco was apparently at the end of its long history.

An American engineer and adventurer gave it a new burst of prosperity. Henry Meiggs conceived the ridiculous idea that he could build a railroad from Callao up to the top of the Andes. Most engineers called him crazy. But he kept at it until he raised $43,000,000 and completed the road. The grade up to the 16,000-foot summit averaged 2.5 per cent, and there were fifty-seven tunnels and many switchbacks. Trains were running over the one hundred and thirty miles to Oroya in 1870, and ever since then the Central Railroad of Peru has been a marvel of railway engineering.

The ancient Peruvian miner or cajonero carried ore in a leather sack on his back and wore a primitive lamp. *(–Cerro Mining Co.)*

With a railway only eighty miles away over the high plateau or pampa, Cerro de Pasco copper became a valuable asset instead of a fatal drawback. The ore ran 25 per cent copper and twenty dollars in silver. The old district soon became prosperous again. Peruvian grandees waxed fat on the profits, and the ever-changing politicians gloated over the easily won graft.

This boom at Cerro was short lived. Two hundred feet below the surface, the mines encountered a great flow of water. The cost of pumping was prohibitive when coal and machinery had to be packed in on llama back. Cerro de Pasco was on the point of dying again. Henry Meiggs tried to come to the rescue with a tunnel several miles long that would drain the mines to a depth far below the bottom levels. The Peruvian government offered him as a reward a quarter of all the metal produced by the mines he drained. Unfortunately Meiggs died when the tunnel was only a thousand feet long. His company fell into the hands of speculators. For many years its only activity was to fleece unwary English and American investors.

The Cerro de Pasco deposits were too great to be long overlooked in the intensive search for copper that took place in the early years of this century. In 1902, engineers sent out by the Hearst-Haggin interests, who owned the big Homestake gold mine in South Dakota, reported that they had found the greatest copper mine in the world. To be sure, it was in the barren Peruvian pampa eighty miles from a railroad and at such an altitude that Americans who tried to live there usually came back bent with rheumatism and emaciated from the hardship of trying to breathe the cold thin air of the Andes. But the engineers were sure the difficulties could be overcome. Haggin investigated and agreed. He convinced J. P. Morgan that the project was feasible. They formed the Cerro de Pasco Mining Company in 1902 and started to shower millions of dollars among the wondering Peruvians.

Geologically, Cerro de Pasco was not a disseminated copper deposit. The ore was not in a great uniform mass from which copper could be readily concentrated. Instead there

The sketch above shows the camp at Oroya, Peru, in the middle of the last century. Below, miners work surface ore about 1902 at the Cerro de Pasco deposit. *(–Cerro Mining Co.)*

were several different grades of copper-silver ore that required different methods of treatment, scattered through a mountain of altered limestone and porphyry. But Cerro resembled the disseminated deposits in the fact that there were many million tons of ore that could only yield a profit if operation were on a gigantic scale. And the investment was just as staggering as that at the Jackling properties.

For five years the construction went on. A hundred and twenty-five miles of mountain railway to Oroya and to the mines that were to supply coal for smelting were the first step in the program. A great smelter nine miles from the mines followed. The American engineers sank deep shafts below the old workings and continued Henry Meiggs' tunnel until it drained the mines to below the bottom of the ore. Dozens of American employees collapsed under the strain of working fourteen thousand feet above sea level. If pneumonia did not kill them, a few weeks' rest among the señoritas and well-stocked bars of Lima usually did. Still the work went on. In 1907 the Cerro de Pasco Company had spent $12,000,000 and was ready to start production.

The copper world trembled at the long advertised flood of metal that was to wreck the market. But only a trickle came. The smelting furnaces did not get enough oxygen in that rarefied atmosphere. They "froze" so solidly that the half-fused ore had to be blasted out. Machinery broke down, and it took three months to replace it. The superintendents at Cerro grew old and gray before their time with the worry. Only superhuman efforts brought production up to twenty or thirty million pounds a year. By 1910 the investment had grown to $23,000,000, and dividends seemed as far away as ever.

Gradually the difficulties were ironed out. Production increased and the cost decreased. Dividends began in 1916. The war years brought all the prosperity that Haggin had hoped for.

After the war, Cerro de Pasco developed other great mines of copper, lead and silver at Morococha, Peru, near Oroya. The silver and lead paid nearly the whole cost, leaving almost

The ancient "Carmen" silver mine at Cerro de Pasco was explored by the Spaniards 300 years ago. A panoramic view of Cobriza Mine, Huancavelita, Peru, is below.

This panorama of the open pit at Cerro de Pasco also shows the new Lourdes Shaft near the top. Below, the old smelter of Cerro de Pasco at Casapalca was operated until 1920. *(—Four photographs, Cerro Mining Co.)*

These are more recent views of the Cerro de Pasco open pit and smelter plant at Oroya. *(—Both, Cerro Mining Co.)*

all the return from copper as profit. In 1923 a new smelter that cost $9,000,000 ended the difficulty with ore treatment. Production climbed to a hundred million pounds of copper a year, and total dividends amounted to more than $60,000,000. Cerro de Pasco failed to wreck the world's copper market as it had threatened in 1907. But in 1934 it was one of the greatest copper mines.

Cerro de Pasco soon had followers that helped spread the fashion of low-grade porphyry copper mines abroad. William Braden was responsible for two of them. Braden had worked for the American Smelting and Refining Company in Montana and was an alert, ambitious engineer. When he heard of Jackling's plans at Utah Copper, he was at once impressed with the great possibilities of the new method of wholesale copper mining. There must be more mines like Utah, and he was going to find one.

Braden had read that the West Coast of South America had produced rich copper ore ever since the time of the Incas. Where there was rich ore, there should be bigger bodies of low-grade ore. While Jackling was still carrying on his experiments, Braden took the steamer to Peru, eager to find the first real South American porphyry copper.

A few months' exploration in the desolate heights of the central Andes took some of the edge off his enthusiasm. Cerro de Pasco had grabbed the only really attractive district in sight. Still South America was a big continent. Farther south, where the Cerro de Pasco engineers had not looked, there might be a better chance. Central Chile was a much pleasanter place to hunt in, anyhow. The beautiful valley around Santiago was like Southern California at its best, and the city itself was the most fascinating on the West Coast. It was Braden's idea of perfect prospecting to make short examining trips up the fifteen thousand foot wall of the snowy Cordillera and in a few hours to be back among the hospitable bright lights of the capital.

He had the happy faculty of making friends with the polite Spanish Americans wherever he went. After a few

The Anaconda party stops in Coro Coro, Bolivia, in 1916; left to right are Reno Sales, L. D. Ricketts, the proprietor of the Gran Hotel, William Braden and Benjamin B. Thayer. Below, Ricketts and the author are on the way to Braden Copper's El Teniente mine. *(—Above, photograph by the author)*

weeks he mastered the art of the back-patting embrace and the constant hand shakes and hat-lifting that mark the Chilean gentlemen. And unlike most American engineers, he learned to speak cultured Spanish. Soon he was "Don Guillermo" to all the best society in Santiago. His Chilean friends entertained him at the Club and the racetrack and in their homes. Best of all they told him all they knew about the old copper mines near their princely estates.

One of these mines was the Teniente. Many years ago the early Spanish explorers had mined rich copper ore under a precipitous ridge of bare red rock that stuck out from the glaciers and snow fields of the high Andes eighty miles south of Santiago. The terrific wintry winds and the constant hazard of avalanches proved too much for the early miners. For a hundred years El Teniente was deserted.

Braden picked a time between snow storms and climbed over the iron-stained cliffs. It was the most wonderful mineralized outcrop he had ever dreamed of. All around the neck of a gigantic dead volcano the rock had been shattered or "brecciated" for a width of many hundred feet. Cementing the fragments was iron oxide that gave the mountainside its brilliant color. And a few feet below the surface the red cementing material changed to copper and iron sulphides. The abandoned ancient workings exposed veins of beautiful rich copper sulphide ore running through the leaner material. They made the whole mass ore of fair grade. The size of it was staggering. Braden's imagination quickly conjured up so many million tons of ore that he could hardly believe it himself. Here was his Utah Copper, but twice as rich. In 1904 he took an option from the Chilean estate that owned the ground and formed the Braden Copper Company to develop his discovery.

The difficulties were just as colossal as the size of the deposit. Only forty miles to the west was the peaceful farming town of Rancagua, on the railroad that led to the port of Valparaiso. But the forty miles were up a cañon where every foot of road had to be blasted out of solid rock. To get electric power for his mine and mill, he must harness the raging

Braden Copper Company's mine office and El Teniente upper tunnel are shown on March 12, 1916. *(—Photograph by the author)*

floods of the Cachopoal River, in a cañon even more forbidding. At the mine itself he must run tunnels to develop the ore. That meant houses for the workmen, perched on a narrow ledge between gulches down which thundered snow slides that often carried away the trails with any unhappy man who was in their way. Far above the snow line, the wintry storms and the avalanches were so terrible that many workmen lost their minds from fear and hardship.

Then Braden had to build a mill to prove that the ore could be concentrated like that at Utah. The only possible place for the mill was on a rocky point down the gulch from the mine and three thousand feet lower, where the snow was not quite so overwhelming. This meant a rope tramway to carry the ore down in buckets slung from steel cables.

Everywhere he turned money vanished by the hundred thousand. And he did not have the money. He had to go back to New York and turn stock salesman, cajoling his friends to advance money against their will for the venture in which they had little faith. The difficulties looked insurmountable. But the money kept dribbling in. And all the time Braden's friends and assistants, Edgar Montandon and Tom Hamilton, kept working on the snaky thirty-inch gauge railway from the mill to Rancagua, and the tunnels and tramway and 250-ton mill. They often had to argue the Chilean workmen into waiting another month before payday came, and they forgot what their own pay checks looked like. But they had such faith in Braden and such an unquestioning loyalty that the work never stopped, no matter how hard pressed they were.

When the end of construction was almost in sight, they nearly lost the whole enterprise. The mill and the mine required a forest of timber. Chilean timber was scarce and expensive. Encouraged by the promise of more stock subscriptions, Braden ordered a shipload of lumber sent down from Puget Sound. His promises of payment when the ship reached Valparaiso induced the lumber company to waive the cash in advance that Braden's standing then demanded. At the last minute the money did not come. The friends who had promised to buy more stock changed their minds. Payrolls were overdue, machinery companies were bringing suit, the lumber schooner was only a few miles from port, and in place of a bank account the Braden Copper Company had a big overdraft. Warrants were out to put Tom Hamilton in jail for fraud. The few assets the company had would be sold to pay a small part of the debts, and that would be the end of the Braden Copper Company.

Just as the final crash was due, the great Valparaiso earthquake of 1906 shook down half the houses on the coast. In an hour, lumber became the most precious material in Chile. All those who had any money wanted to rebuild at once, at any cost. And the only chance to do it quickly was with frame buildings. Every stick of timber within reach was sold in a few days. At the moment when the shortage was most acute, in sailed Braden's ship loaded with millions of feet of the best Oregon pine.

It went like hot cakes, at prices that made even Tom Hamilton ashamed. The mill could wait when money was coming in like that. Before the ship was empty the Braden Copper Company was out of debt, with all the money it needed in the bank.

The worst troubles were over then. There were lots of engineering difficulties, but hard work would solve them. Late in 1906 the 250-ton mill started, and soon proved that it could make a rich concentrate. Four million tons of 3 per cent ore were already in sight. Every round of shots in the tunnel added to the reserves. It would still take millions to build a big mill and a smelter, but success was assured.

Braden could afford to wait now until he could get his own terms for additional financing. In 1909 his old employers, the Guggenheims, bought control of the Braden Copper Company. Braden himself received a big cash payment and kept enough stock to give him the fortune his efforts and worry deserved. And all those who had stayed with him so loyally when the venture looked hopeless got their share in the reward.

The Guggenheims and their successors, the Kennecott Copper Corporation, poured in $25,000,000 before they made the Braden a real success. First came a three-thousand-ton mill and a smelter. The mill was too crowded down in the narrow gulch and recovery was far from satisfactory. Before long they built another larger mill in a wide spot farther down the cañon. It was ten years before the difficulties were finally overcome. The accounts showed an operating profit, but every cent of it went back into new construction or bond interest until 1923. Luckily the Kennecott Corporation could afford to wait for its profits. They knew they had one of the most valuable mines ever developed. Every year the ore reserves grew by many times the amount of ore mined. And the ore was so rich that Braden finally became one of the lowest-cost copper mines in the world. The Braden Copper Company turned out 64,000,000 pounds of copper in 1917, and 200,000,000 pounds in 1927. Even at this rate the 240,000,000 tons of 2.2 per cent ore would last fifty years. Braden was almost equal to Utah in the amount of copper it had ready to mine. Together with Utah it ensured the Kennecott Copper Corporation, which now owned them both, a leading place in both foreign and domestic copper production for a long generation.

Cerro de Pasco and Braden just gave the flood of South American copper a good start. The greatest of all copper mines was waiting for someone with courage and vision enough to lift it from obscurity.

Chuquicamata had been a well-known copper district for many years. Ever since the Spaniards came to the Americas, the rich stringers of blue copper sulphate ore in this bare ridge near the north edge of the Desert of Atacama had yielded all the copper the Chilean miners could sell. For many years the copper was packed on llamas to the coast, ninety miles west and nine thousand feet down. Then the two and a half foot gauge railroad was built from the port of Antofagasta to Bolivia, passing within a couple of miles of the Chuquicamata mines. As early as 1903, an English and Chilean company planned a branch railroad from the mine to the main line at

Reno H. Sales, left, and the author pose aboard their ship from Buenos Aires to New York, 1916.

Calama. There they could develop enough water for a small smelter. Production grew to 3,000,000 pounds a year, to the joy and pride of the English owners.

Mining at Chuquicamata had gone on for so many years that the long ridge was like a prairie-dog town, with its hundreds of old shallow workings. But a prairie-dog would soon starve in the desert of Atacama. In that waste of yellow rock and sand the natives say it "looks like rain every seven years." There was not a bush or a blade of grass to relieve the repelling beauty of the desert. For many years the few American engineers who visited "Chuqui" saw no possibility of operating the mines on a larger scale. They were glad to come away as soon as they could from the bitterly cold wind that constantly blew down from the snowy heights of the Andes.

Occasional visitors did pause long enough to wonder what lay between the streaks of rich ore in the old workings. As early as 1903 a metal dealer and engineer named Duval wrote to Daniel Guggenheim that there was a great copper ore body at Chuquicamata. The Guggenheims decided that the problem of devising a cheap method of recovering copper from the sulphate ore, different from any other known ore body, was too expensive and uncertain to justify an undertaking in such a forbidding locality. Eight years later Louis Ross of Boston visited the district. Ross saw wonderful possibilities for mining the enormous mass of leaner ore between the rich streaks. He returned to Boston and told A. C. Burrage of the Amalgamated group of copper magnates about his theories. Some months earlier Claude Vautin, the engineer for the Calama Company, had cabled Burrage about the possibilities at Chuquicamata. The question of how Burrage first heard of the district led to a long lawsuit brought by Ross, who demanded a commission of many million dollars. Apparently Ross and Vautin together persuaded Burrage that the ore body was worth developing. Vautin secured options on most of the properties in the district, and Burrage went to the Guggenheim Brothers to secure help in the financing.

Outcrop and old shallow workings at Chuquicamata, the greatest copper deposit in the world, are shown above in 1916 when Chile Copper Company (Guggenheim Exploration) started open-pit mining. The open pit in 1963 is below. This deposit contained more copper than all the mines of Zambia (Northern Rhodesia) combined. Its maximum production was 312,000 tons of copper from more than 20 million tons of ore in 1969, just before Chile's Marxist government took the property and plant away from Anaconda. *(–Above, photograph by the author; below, Anaconda Annual Report)*

Pope Yeatman was then in charge of operations at the Braden Copper Company for the Guggenheims. His engineers and scouts were looking for other mines. They heard of Chuquicamata after Burrage had secured the options. Yeatman had learned by experience how costly it was to equip great mines in the Chilean Andes. His company was the only one in the country with sufficient resources to tackle the job. Even if others had secured options, they must eventually come to the Guggenheims for money. In order to be ready when the call came, Yeatman sent his examining engineers, Fritz Mella and Robert Marsh, to Chuquicamata. They reported that the whole mountain, eight thousand feet long and three thousand feet wide, was made up of copper sulphate ore averaging 2 to 3 per cent copper. The quantity of ore was practically unlimited, but the treatment of it was a difficult problem. They sent large samples to the Braden laboratory for testing. By the time Burrage asked the Guggenheims to join him in the enterprise, Yeatman was ready to cable a glowing report on the mine and to state how the copper could be recovered.

In 1912 the Guggenheims and Burrage formed the Chile Exploration Company and went to work. The usual years of drilling and of metallurgical tests followed. Pope Yeatman had been so successful at Braden that the Guggenheims put him in charge at Chile as consulting engineer. The best metallurgist they could find – E. A. Cappelan-Smith – built test plants in New Jersey to treat the ores sent up from the mine. Gradually the ore body grew to astounding size. The drill holes proved that the whole mountain was ore. The engineers estimated a hundred million tons, then two hundred, four hundred, finally six hundred million tons of ore assaying more than 2 per cent copper. Meanwhile Cappelan-Smith perfected the new process of dissolving or leaching out the copper from the ore in huge vats and precipitating it by electricity. The ore itself, oxidized in the extremely arid climate of the Atacama desert, contained enough sulphuric acid to dissolve its own copper. All the metallurgists had to do was to start the process going and to keep loading and emptying the vats at the right time.

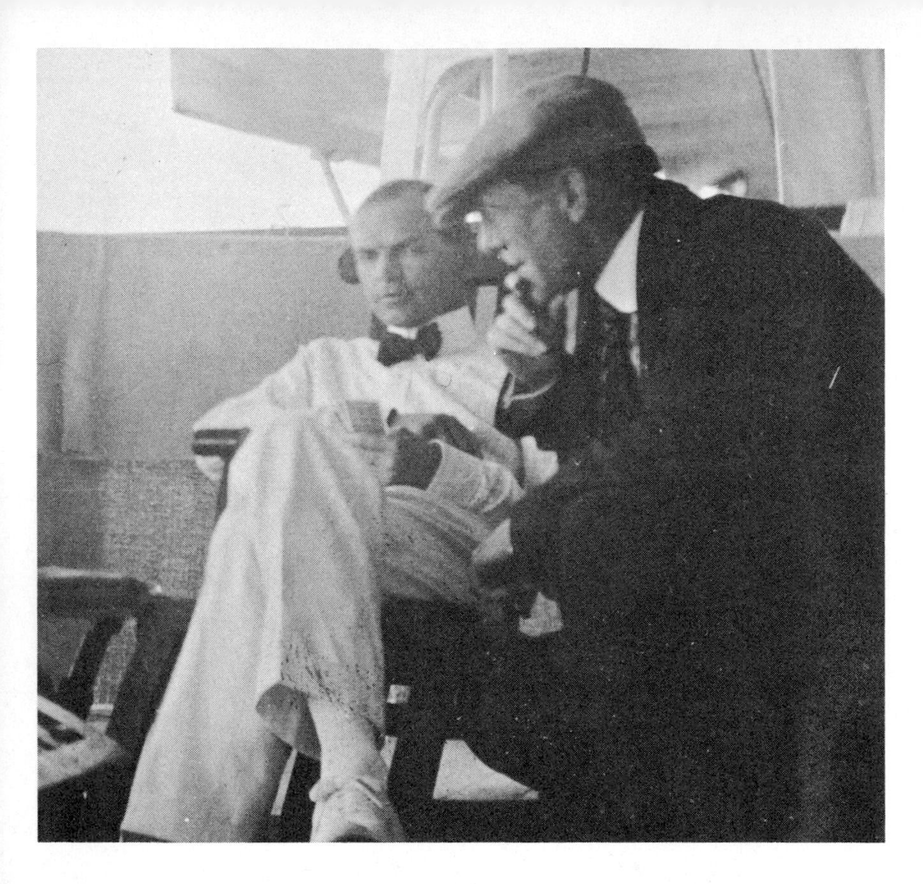

Above, Sales and Ricketts are on board the Chilean steamer *Limari* in 1916. Below are the Potrerillos camp and outcrop of Andes Copper, also in 1916. *(—Both, photographs by the author)*

By the end of 1913 they were ready to start building the big plant. It was the greatest copper ore body in the world, and the scale of operations must be in keeping with the developed tonnage. Five thousand tons a day was the first objective. This was raised to ten thousand, and soon an increase to sixteen thousand tons a day was planned. An operation of this bewildering size required money on a scale that even Jackling never dreamed of. A hundred million dollars was poured into the desert before Chile Copper was a success. Great steam-shovels to mine the ore; a whole army of leaching tanks that held ten thousand tons each; powerful machines to grind the ore and to load and unload it in the tanks; electrolytic-precipitation apparatus on an unparalleled scale; a seventy-mile pipe line from the nearest stream up in the Andes; powerful steam turbines driving generators of many thousand kilowatts' capacity, with a transmission line from the nearest seaport to the mine; towns, offices, all the other items in a new mining enterprise were on the greatest scale ever seen. And in that remote corner of the world they cost twice as much as they would have cost at home.

In 1915 Chile Copper was at last ready to start production. But the troubles had just begun. Someone had forgotten that the power-plant must use sea water to cool its condensers. The salt ate out a million dollars' worth of machinery in a few months. The magnetite anodes that worked so well in New Jersey did not like the Chilean climate, and soon made a half-million-dollar pile on the scrap heap. With the new process, unusual ore, and the great distance from supplies, all the difficulties were so magnified that there seemed to be no end to the spending. Yeatman and Cappelan-Smith and their assistants kept at it until one by one the problems were solved. By 1916 Chile Copper was making a large paper profit. Dividends were still far in the future. The size of the ore body demanded greater and greater productive capacity. For eight years, all the money the mine made went back into plant enlargements. The copper world knew that Chuquicamata was the greatest of all copper mines. And the engineering profession gave credit to Yeatman and Cappelan-Smith and Fred Hellman and the others for overcoming all the difficulties. But until 1923, ten years after construction started, the stockholders did not receive a dollar in dividends.

Chile did not long remain a Guggenheim property. After the war, the Anaconda directors realized that Butte had lost its position as the leading copper district in the world. The mines were four thousand feet deep, and costs were mounting. Anaconda had paid a great price for the American Brass Company, in order to avoid having to pay others for fabricating its copper. Now it was threatened with the prospect of having to buy copper for American Brass from other mines, at fancy prices. Unless it could acquire new mines even greater than the old ones Anaconda seemed destined to decline until it became a second-rate company.

It was hopeless to try to find mines great enough to fill the requirements. Only one mine in the world would do that. This mine was Chile Copper. The Guggenheims did not want to sell. Anaconda raised the bid until it was impossible to refuse. In 1923 the Guggenheims finally accepted an offer of $35.00 a share for their controlling interest in Chile Copper. This meant a valuation for Chile stock and bonds of more than $180,000,000.

The Anaconda directors doubled the productive capacity again, largely by technical improvements that required a comparatively small investment. In 1929, Chile Copper made 300,000,000 pounds of copper, and it could have made 75,000,000 pounds more. The capacity can be doubled again at moderate cost. The only limit to the output of Chile Copper is the amount of copper it can sell.

Chile's ore reserves, exceeding 600,000,000 tons, were already far greater than any other mine could boast. Still the Anaconda directors were not satisfied. They sent Reno Sales, their chief geologist in Butte, to see if there was likely to be any more ore at Chuquicamata. Sales conceived the idea that at the north end of the mineralized ridge the ore in depth might turn to the west along a cross fault and become richer.

He planned drill holes to prove his theory. The result was 400,000,000 additional tons of ore much higher grade than that found before. Chile now has developed the stupefying amount of a billion tons of ore that assays nearly 3 per cent copper. Nothing like this has been dreamed of in mining history. The Chile Copper ore body contains more than twice as much developed copper as all the mines in the United States, in ore more than twice as rich. Chile has ready to mine two-thirds as much copper as the whole world has used since men were apes. It has much more ore and more copper developed than all the widely advertised South African mines put together, and in the long run its cost will be lower than theirs. Anaconda paid a high price for Chile. By doing it, she acquired a mine so much greater and richer than any other possible mine or group of mines that for generations Anaconda must be the controlling factor in the copper business of the world.

One more mine completes the list of the porphyry coppers. This is the Andes Copper. William Braden was the father of the Andes. When he sold the Braden Copper Company to the Guggenheims, Braden was too ambitious to sit down and enjoy his profits. He had built up a wider acquaintanceship than that of any other American on the West Coast, and everyone liked and trusted him. His Chilean friends were constantly telling him of properties they thought promising. If there were any other great mines in Chile, Braden was the one to find them.

The Anaconda Company had formed an alliance with Braden in 1913, and made him its South American representative. He soon had his old assistant, Tom Hamilton, and a corps of engineers scouring the Andes for new mines that might become great. They studied and rejected hundreds of prospects. One seemed to meet the requirements. In a barren valley near the southern end of the Atacama desert, ninety miles from the port of Chañaral, the Chileans had mined stringers of rich copper ore. They called the camp Potrerillos or "little fields," from the pitiful small patch of grass and brush that crowded close to the oozing springs in the lower end of the valley, the only green spot for many miles around. Surrounding the shallow workings, Braden's engineers found hundreds of acres of yellow and brown and green stained porphyry. This might be the outcrop of another disseminated ore body.

With Anaconda money Braden built a road through rock and sand up the fifty-five mile cañon from the end of the narrow-gauge railroad that ran inland from Chañaral. He sent in churn drills and started to develop the ore. And he found an ore body that would have satisfied the most grasping copper magnate if it had not been overshadowed by the far greater deposits of the Chile Copper and the Braden Copper Companies. Potrerillos developed 130,000,000 tons of 1.5 per cent ore. The usual expensive campaign of construction followed. Water for the mill was almost as difficult to obtain as at Chuquicamata. The nearest stream large enough to supply the necessary ten thousand gallons a minute was at La Ola, in the high Andes thirty-five miles to the east. The pipe line from La Ola cost several hundred thousand dollars. Just as the line was finished, a deep extraction tunnel under the Potrerillos ore body broke into an underground water channel that no one had suspected. A torrent of many thousand gallons a minute drove all the workmen out of the tunnel for months. For a time it looked as though the pipe line from La Ola had been wasted. It was several years before the underground water drained off and the pipe line was needed.

The Potrerillos ore was partly sulphate and partly sulphide. Owing to the irregular occurrence, both types of ore must be mined at once. This meant a complicated metallurgical problem. Fred Laist, the brilliant metallurgical manager of the Anaconda Company, devised a process of treatment that would yield cheaper copper than the method that had been used on simpler ores. As the equipment would cost too many millions even for the Anaconda treasury, Braden and the Anaconda interests formed a subsidiary called Andes Copper Company and sold enough stock and bonds to the public to complete the $50,000,000 total expenditure. After many delays the railroad, mine equipment, flotation mill, leaching plant and smelter were ready to operate. The remote location caused great delays and added expense as it always does. Andes was not ready to produce copper until the end of 1926, twelve years after the ore body was developed. In 1929 it treated 6,000,000 tons of ore and made 162,000,000 pounds of cheap copper. Then the Depression came along and forced curtailment to a fifth of capacity just as the new plant was beginning to reap the reward of all the hard work and the huge investment. Andes was the last of the porphyry coppers and the unluckiest of them.

AUTHOR'S NOTE: After *Romantic Copper* was published, Mr. Pope Yeatman, who had been in charge of Chuquicamata and other Chilean properties for the Guggenheim Exploration Company, wrote me a friendly letter enclosing five pages of corrections that he thought should be made in my story of Chuquicamata. Most of the suggested changes concerned technical questions. Mr. Yeatman says he heard of Chuquicamata before Burrage had optioned it. Yeatman was negotiating for part of the property when he was told of the Burrage option on 40% of the area. After the Guggenheim Company acquired the property, the amount raised totalled fifty instead of a hundred millions, as stated in *Romantic Copper*. Profits turned back into the plant to increase the investment added to the "book" outlay.

Delay and added cost in building the Chuquicamata plant were due in part to the fact that World War I prevented shipments of machinery from Germany. The war also cut off shipments of magnetite anodes from Germany; Messrs. Yeatman and Cappelan-Smith knew that the substitute alloys would not be as good, but nothing else was available. Difficulty with the condenser tubes due to using sea water was easily overcome while the plant was in operation.

Finally Mr. Yeatman says that the Chile Copper plant was working satisfactorily before that at Ajo, which copied much of the Chile design.

It is too bad there were differences of opinion about the early operations at Chile. However, I am glad to give Mr. Yeatman's criticisms of the version that I obtained from other sources.

The statistics given in the above pages on the disseminated copper deposits are those in the original *Romantic Copper*, written in 1931 to 1933. All of the deposits have grown much greater since then, largely through including much leaner material in reserves. There have been so many consolidations and changes in ownership since 1933 that figures on additional profits and even tonnages of ore treated at the various mines are seldom available. They are so great as to be almost astronomical.

PRODUCTION AND CONSUMPTION IN BALANCE

Within six years after Jackling started the Utah Copper mill, all of the great disseminated copper ore bodies had been discovered. They brought a flood of new production that no one had dreamed possible. The fear of a copper famine that was so imminent in 1907 quickly gave place to an equally disturbing fear of overproduction. The world output of copper jumped to nearly two billion pounds in 1911, and the price fell from twenty cents to twelve cents. The business revival of 1912 and 1913 put copper up to sixteen cents again, but only for two years. In the early part of 1914 the price dropped once more. Copper experts agreed that consumption could not possibly keep up with the rapidly increasing production. The copper industry was evidently in for a long period of low prices, like the one that followed the development of Butte, and of the southwestern enriched deposits in the 'eighties and 'nineties.

The World War changed the situation completely. Munitions factories were soon crying for copper. By 1916 consumption jumped to three billion pounds a year. If Jackling had not brought porphyry coppers into style just when he did, the Allies would have suffered from as acute a copper shortage as did Germany. This might have brought victory to the Central Powers. Even with the great concentrators at Utah, Nevada, Braden, and all the other porphyry properties running full blast, there was hardly enough copper for the shells and new electrical equipment and automobiles and all the other war uses. The price climbed to thirty-four cents a pound in 1916. The following year the United States Government fixed the domestic price at twenty-three cents, and the world followed suit. Even at this enforced low price the copper companies wallowed in dividends. The universal labor shortage prevented them from increasing production as much as they wanted to, but at three billion pounds a year there was all the profit they could reasonably ask for.

Then came the end of the war, and chaos for the copper producers. During the first few months after the armistice the copper miners were urged by the War Department to keep up production on the chance that the war might start again. They did so, but no one wanted the copper. By 1921 nearly a year's supply of unsold metal was piled up at the smelters and refineries. The consumption had fallen to less than half the wartime rate, in spite of a drop in price to twelve and a half cents a pound. Desperate measures were necessary. Nearly all the great American mines shut down completely, and their foreign competitors cut production 40 per cent. Even at that, few expected that the accumulated copper surplus would be used up within three or four years. The industry was sunk in a slough of despond.

If new copper districts had been found and brought to production at this time, the situation would really have been hopeless. But Fate laid her plans as skilfully as she had always done when an oversupply of copper threatened. In the eleven years from 1914 to 1925, not a single great new copper mine was developed in all the world. Mining and exploration companies spent millions of dollars hunting for new ore bodies. The only result of the search was to add to the reserves of the porphyries a few hundred million tons of marginal material that averaged less than 1 per cent copper. Improving metallurgy and cheaper mining led to the hope that this low-grade ore might some day yield a profit.

The low metal price in the years that followed the war proved to be a blessing to the copper industry. It stimulated the use of copper beyond the wildest expectations. In addition to the electrical equipment that continued to take over half the production, many new fields of consumption were opened up. The Copper and Brass Research Association, representing the largest copper companies, carried out a successful advertising campaign. It taught architects to specify copper or brass pipe and copper roofs and proved that for a hundred other uses copper was far superior to any other material. And at twelve or thirteen cents a pound, the Research

Association said, the one who bought copper articles could be sure of a scrap value nearly equal to the price he paid. The argument was most convincing.

The result of this advertising and of the steady growth of industry after the 1922 depression was a geometrical increase of 10 per cent a year in the copper consumption of the world from 1920 to 1925. The unsold surplus that stifled the industry in 1922 was all gone two years later. The shutdown of the great American companies lasted for only a few months. Then all the copper mines in the world started to produce at a rate that soon exceeded the wartime output. In the years from 1923 to 1928, the price of copper remained stable at thirteen to fourteen and a half cents. This price encouraged consumers to use copper rather than other metals, and at the same time it gave the great companies a margin of profit that made their mouths water. The lot of the copper magnates had never seemed so rosy.

The demand for copper was growing so rapidly that by 1925 the specter of a copper famine reared its head once more. At the 10 per cent yearly rate of increase, even the great ore reserves of the porphyry coppers could not long keep up the supply. A few minutes' work with a slide rule proved to the more optimistic copper engineers that if consumption continued to increase 10 per cent every year, even if the world found as much new copper as it had developed in all of history, within forty years every copper mine in the world would be worked out. That would mean the end of electricity and of our industrial civilization.

There were a lot of "ifs" in this reasoning. In spite of them, it was evident that the world needed new copper deposits. And right on time Fate presented us with a great group of ore bodies of an entirely new style – the African copper deposits.

THE COPPER OF AFRICA

It is typical of Nordic conceit to call the African deposits new. They were new to us, but they had been well known to the black tribes of the Congo and Zambesi watersheds for hundreds of years. When Stanley made his first adventurous journey across Darkest Africa in 1873 to 1875, he found the savage chiefs and their shining black wives wearing leg bands of beaten copper wire. Copper rings and spears and other utensils were in common use. Later explorers saw the natives mining rich oxidized copper ore in open pits, and melting it down in little furnaces built of the clay of termite nests. The process was laborious, but in the unrecorded centuries the primitive smelters had treated an astonishing amount of ore. In the Belgian Congo alone there are old workings in more than a hundred separate copper ore bodies. One of the pits the white men found was seven hundred feet long, four hundred feet wide and thirty feet deep. The natives had carried out on their backs half a million tons of the soft surface ore from this mine alone.

Stanley was the first to recognize the mineral wealth as well as the trading possibilities of Central Africa. Soon after his first expedition, he induced King Leopold II of Belgium to back further explorations. By agreement between the European powers, the watershed of the Congo was given to Belgium as a sphere of influence and later a colony. Leopold horrified the world by his heartless treatment of the natives in the ivory and slave trades, but failed to follow Stanley's suggestion to develop the mineral resources of the Congo.

Meanwhile Cecil Rhodes was extending the British sphere of influence northward. Though the English government had given his British South Africa Company complete power over the thousands of square miles later called Rhodesia, lying between the Transvaal and the Congo, he was not satisfied. His scouts told him the richest mineral area was farther north, in the Belgian territory. The Cape to Cairo railway, already built fifteen hundred miles north from Cape Town, would soon make this remote region accessible. The British South Africa Company ought to have a concession from Leopold granting it exclusive right to explore the upper Congo basin.

Rhodes went to King Leopold in 1890 and found he was too late. The King had already given the concession to the

Countess of Warwick, who had no idea of releasing it for any paltry consideration. Rhodes succeeded in buying her out, but on almost prohibitive terms. The British South Africa Company was to do all the work and spend all the money, while the Belgian government retained the right to take a 60 per cent interest in any successful venture.

Cecil Rhodes himself was so busily engaged in the political struggle between Cape Colony and the Boers that he could not develop the new concession. His most trusted lieutenant was Robert Williams, an English engineer and soldier of fortune nearly as remarkable as Rhodes himself. Rhodes therefore gave Williams the job of developing the mineral deposits in Northern Rhodesia and in the Congo.

Williams lost no time. In 1891 he sent George Grey to the Congo to hunt for the old copper mines that Stanley had told about. Grey and his party made the long journey from the end of the Cape railway in southern Rhodesia on foot aided by a party of sixty native bearers. The journey in and out took many months. About all the bearers could do was to carry their own food. When they finally reached the Congo, supplies were already so low that they could only prospect for a few days before starting home again. In this short time they found a hundred and fifty copper prospects in a two-hundred-mile strip of tropical jungle and grassy plateau called Katanga. The copper zone was just north of the Rhodesian border and southwest of Lake Tanganyika. Among the old mines was the great seven-hundred-foot open cut that Grey named the Star of the Congo.

Until new railways were built, it was hopeless to try to do anything with the Katanga mines. To Lobita Bay, the nearest seaport on the west coast of Africa, the thirteen-hundred-mile journey down the sluggish tributaries of the Congo and through the fever-infested jungle took several weeks. Beira, on the east coast, was seventeen hundred miles away through country nearly as unhealthy and dangerous. The only route that was even moderately safe was the two thousand-mile one from Cape Town north through the diamond fields of the

Transvaal and through Rhodesia. The end of the railway from Cape Town was less than five hundred miles from the Star of the Congo. But the journey from the railway was still a hazardous one. On the rolling plateau, dense forests were broken by open spaces covered with shoulder-high grass, in which lurked lions and leopards watching to prey on the giraffes and the dozen varieties of antelope. Great pythons laid in wait to crush any unfortunate man or animal that came within reach. The tsetse fly with its terrible follower, sleeping sickness, killed or drove away all the men and horned animals in areas of hundreds of square miles. In the marshes along the rivers, myriads of mosquitoes were ready to infect the prospectors with malaria. Black-water fever was a still more deadly threat. There were no roads, and in the summer months torrential rains changed the footpaths into morasses of black mud. The scattered villages of fierce native tribes often attacked the prospectors who ventured into their territory without adequate guards. The bickerings and finally the war between the English and the Boers shut off communication between Rhodesia and Cape Colony for months at a time. Under these circumstances, development of the great concessions held by the British South Africa Company was slow. Railroad building was even slower. For nearly ten years Rhodes and Williams made no attempt to work their new copper deposits.

In 1899 Williams was at last ready to act. He secured from the British South Africa Company any rights that remained under the old agreement with Leopold, together with a large concession in Northern Rhodesia. Tanganyika Concessions Ltd. was the name of the company Williams formed. In its name he made a new prospecting agreement with the Belgian government. Tanganyika was given a monopoly on all mining in an area of 60,000 square miles in the upper Congo basin. In return Belgium was to have a 60 per cent interest in any mines developed.

Then began Robert Williams' twenty-year fight against the jungle. Development soon proved to his satisfaction that

Bwana M'Kubwa, near Ndola, is the oldest of Zambian (Northern Rhodesian) copper mines. As shown above, it was shut down for many years due to metallurgical difficulties; production started again in May, 1971. Below, the Kansanshi Mine is a Zambian deposit on which work is said to be planned. *(—Both, Charter Consolidated Ltd.)*

Katanga contained several of the richest and greatest copper deposits in the world. In the Star of the Congo and the Kambove alone he found 10,000,000 tons of 12 per cent copper ore. In a reasonably accessible country, these mines would have been bonanzas. In the heart of equatorial Africa, it seemed for many years a hopeless task to make a profit out of any copper ore bodies, however rich.

Delays and disappointments could not discourage Williams and his chief assistant, George Grey. Year after year they came to the bankers of London with tales ever more glowing, and with demands for still more money. The financing became so involved that it was almost impossible to say who owned what. A Belgian company called Union Minière de Haut Katanga was formed in 1906 to hold title to the mines and to operate them. The Belgium Government controlled Union Minière, with Tanganyika Concessions as a large minority stockholder. Both companies sold common stock, and preferential stock, and debentures, and bonds, and every other known form of security. Every year Williams' promises were more seductive, and every year the fulfilment was postponed. Many of the leaders in the mining world began to doubt if the ore was there at all. Maybe Williams was just a skillful promoter, who had picked a mine so far away that it would cost thousands of dollars to send anyone to check his claims. As late as 1911 the usually dependable *Copper Handbook* reflected this suspicion in its sarcastic references to certain reports concerning the results of exploration of one of the Katanga ore bodies. Its comments clearly showed that it had no confidence either in the mine or in the method of financing that had been pursued.

Fever and lethargy and the moral breakdown that overcomes so many white men in the tropics were even more serious obstacles than lack of money and inaccessibility. A constant stream of ambitious young Belgian, English and American engineers and artisans went out to the Congo to make their fortunes. Many of them stayed in lonely graves in the jungle. Hundreds more came back physical and moral wrecks. Constant association with the thousands of black laborers who were little more than animals took away all pride in humanity. The afternoon glass of whisky that washed down the daily dose of quinine that was a necessary precaution against malaria — the "sundowner" of the British colonies — often started a nightly debauch. Katanga was not an encouraging place for one who had faith in mankind.

Even the beasts shared in the conspiracy to wreck Williams' plans. One day his most valued assistant, George Grey, was looking at a new prospect. A lion suddenly sprang from the tall grass. Grey died the death of a hunted animal. It was no wonder that so many of the European employees drank heavily to forget the menace of the forest that surrounded them.

On every side bad luck followed the venture. Williams fought his way to success in spite of it. By 1910 he completed the two hundred miles of railway to connect with the long line to Cape Town. At last he could bring in equipment for a smelter at the Star of the Congo. The smelter started in 1911. But the troubles were just beginning. Mining was simple enough. The Katanga ore occurred in great lenses impregnating beds of impure sandstone. The Star of the Congo body was two thousand feet long and ninety-six feet wide, and the Kambove three thousand feet long and two hundred feet wide. As there was only a shallow covering of soil, mining by steam shovel cost only a few cents a ton. Unfortunately the treatment of the ore was more difficult. Instead of a sulphide that could be concentrated before smelting, Kantanga copper was in an earthy, oxidized mass that must be smelted directly. It contained so high a percentage of silica that the smelter froze repeatedly. Even when it kept running, the sticky slags contained 4 per cent copper. After thirteen years of unfulfilled promises, Robert Williams had spent $33,000,000, and the best he could show for it was a million pounds of copper a month that cost half again as much as it was worth. No wonder the stockholders and engineering critics were skeptical.

Gradually the production grew, and the cost of copper came down. In 1917 the Katanga output was 66,000,000 pounds, and the cost was well under the wartime price. Union Minière paid 22,000,000 francs in dividends in 1918. After a struggle that had lasted nearly twenty years it had at last become a real factor in the copper industry.

The taste of success only whetted the appetite of Robert Williams and his associates. They had boasted for twenty years that they had the richest copper mine in the world. Now they were determined to make good the boast. The 12 and 15 per cent ore was nearly gone. In place of it Union Minière had developed 75,000,000 tons of 7 per cent ore. This was still remarkably rich, in spite of the difficult metallurgy. But it required a new smelter, an enlarged leaching plant and the completion of the railway to the West Coast before Katanga could become a low-cost copper producer. Williams discontinued the dividends that had barely begun, and put all the profits back into new construction. Production climbed steadily until in 1930 it reached 300,000,000 pounds. The American copper magnates at last realized that they had a competitor strong enough to wreck their plans for controlling the metal price through an agreement to limit production. For a few years Katanga became the bugaboo of the industry.

If it had been Katanga alone, the threat would not have been serious. Great as the mines of the Belgian Congo were, they could at best supply less than 10 per cent of the copper the world needed. But Union Minière started an entirely new style of copper ore bodies. Fate and the American prospectors were slow to follow the fashion of bedded ore bodies in sandstone and shale. When they finally came to it they banished any fear of a copper shortage for many years by the most spectacular discoveries that copper had ever known.

South of Katanga, only a low divide between the Congo and the Zambesi river basins marks the boundary between Belgian territory and the British Crown Colony of Northern Rhodesia. The same rocks and the same geological structure

This rig is drilling for ore on the Northern Rhodesian copper deposits in 1929. *(—Roan Consolidated Mines Ltd.)*

occur on both sides of the border. Why should all the copper mines be in Congo territory?

The Tanganyika Concessions engineers asked this question at the beginning of this century. They even went further and found the copper. In 1908 they started a smelter at the Kansanshi Mine; the ore body proved to be small and lean, and the enterprise was a failure. Other companies acquired concessions from the British South Africa Company and developed ore bodies like those of the Katanga at Bwana M'Kubwa and at Nchanga. Expensive experiments and small leaching plants proved that the oxidized ore was just enough leaner than that at Katanga to take away all hope of profit. By the beginning of 1925, the copper companies of Northern Rhodesia were in a pretty hopeless condition.

Among the companies that had tried to find a new Star of the Congo south of the Rhodesian border was Selection Trust, Ltd. A distinguished American who was then living in London was the leading figure in Selection Trust. A. Chester Beatty had been a close associate of Herbert Hoover and of John Hays Hammond and the other American engineers who had achieved great success by combining engineering with promotion. Beatty made an alliance with Edmund Davis, one of the shrewdest financiers and promoters in London and Africa. They floated half a dozen development companies with interlocking directorates and began operations at various points. Rhodesian Congo Border Concessions developed the Nchanga mine and shut it down as too lean to work. Bwana M'Kubwa Copper Mines Ltd. failed with the original Bwana M'Kubwa mine and shut down a new prospect called the Nkana without serious development. Selection Trust held most of the remaining country along the Congo border.

In 1925 Beatty and their associates knew that there was copper in the Selection Trust area. They were very doubtful if it had any value. Except for two diamond drill holes at Nchanga, that might be freaks, all the Rhodesian ore was oxidized like the Katanga ore. But the grade was only 3 or 4 per cent instead of 7 per cent. If Katanga had difficulty

Above are Sir Alfred Chester Beatty, left, and William Collier. Below, a MacNamara shot drill is being set up for exploratory drilling in Zambia in 1929. *(–Three photographs, Roan Consolidated Mines Ltd.)*

in making ten-cent copper from 7 per cent ore, what chance had they with 3 per cent ore of the same refractory character?

Still there was a bare chance that at greater depth the ore would change to a sulphide that could be cheaply concentrated before smelting. Beatty's geologist, R. J. Parker, thought it was worth a try. Selection Trust sent him to Rhodesia to study the possibility more carefully. Parker reported that the chance looked good enough to warrant a little development on the Roan Antelope.

William Collier had found the Roan Antelope twenty years earlier. While on a hunting and prospecting trip through the wilderness near the great Victoria Falls of the Zambesi he saw an old medicine man in a Bantu village using powdered malachite as a charm. Collier knew that malachite was a rich copper ore. He bribed the medicine man to tell where it came from. Following the crooked Luanshya River, he came at dusk to a horseshoe-shaped clear space, or "damba," in the forest. Across the clearing he saw a roan antelope grazing. This meant meat for the native bearers. A rifle shot brought the antelope down. Stooping over it, Collier saw a piece of rich malachite ore. Later he found grass-covered pits dug by the natives. He located the ground as the Roan Antelope claim. No development worth noting was done until twenty years later, when the ground was included in the Selection Trust Concession and was chosen by Parker as the site for his first prospect shaft.

The advance of civilization toward the Congo had left the Roan Antelope in much the same condition as when Cecil Rhodes made his first trip to the Zambesi in the 'seventies. Game of all sorts was too abundant for comfort. One day when a black workman climbed out of the shaft he looked over the edge and found himself face to face with a lion. He quickly retreated down the ladder and stayed "treed" in the bottom of the hole until the next day, when the lion wandered away. Parker did much of his prospecting with a pick in one hand and a rifle in the other.

Above are Thorold F. Field and Russell Parker, right. Below, the 1928 Chevrolet 1½-ton truck was invaluable for moving materials in the almost trackless Zambian wilderness. *(—Three photographs, Roan Consolidated Mines Ltd.)*

In December, 1925, Parker cabled Beatty that two of his shafts had entered 3.5 per cent disseminated chalcocite ore. This sulphide could easily be concentrated. Still Beatty and his associates were lukewarm about the venture. The Roan Antelope was many thousand miles from London, and it would take hundreds of thousands of dollars to prove whether 3.5 per cent ore in Northern Rhodesia had any value at all.

Two or three months later, an American engineer named T. F. Field appeared in London looking for information about Rhodesia. His New York clients had a theory that the next great mining development would be in central Africa. Beatty was glad to tell him all about the Roan Antelope and to give him an option on a large block of stock in a subsidiary company that Selection Trust would form to develop the mine.

In April, 1926, Field and Beatty's associate Selkirk reached the Roan Antelope. The trip had been a long and hard one. In Southern Rhodesia they saw only small ore bodies of insignificant value. Farther north Bwana M'Kubwa and Nchanga had only oxidized ore so much leaner than that at the Katanga mines that it seemed a hopeless task to make money out of it. The rainy season was just over, and it was hot and muddy and they were thoroughly tired and disgusted. They were all ready to turn the Roan Antelope down at first sight and to hurry back to the comforts of London.

Parker met them at the railroad and drove them over the rough road to the prospect in a rattle-trap car that did not add to their optimism. He told them of his sulphide ore, but made little impression on their gloom. At last they stood on the dump of the prospect shaft. The rock looked like ordinary barren shale. In complete disgust Field picked up a piece and looked at it with his pocket lens. He almost shouted in amazement: "Why, it *is* chalcocite."

All through the shale were fine specks of the rich copper sulphide. There was practically no other metallic mineral. Field saw that the ore could be easily concentrated into a very high-grade smelting product. If only there were enough of the ore, this would mean wonderfully cheap copper.

The Kambove, in Katanga (now Zaire), has reached final depth in the south pit, shown above; further mining is underground. Below are headframe, bin and cars of the old Kansanshi Mine. *(—Above,* WORLD MINING *and Miller Freeman Publications; below, Charter Consolidated Ltd.)*

There was no more weariness or pessimism in the party. Parker had worked out the geology so well that in a few hours he showed them the bed of copper ore twenty to thirty feet wide, outcropping in the treeless "damba" around the nose of a blunt canoe-shaped fold in the shale and sandstone. From the point of the curve he had traced the ore for a mile along both sides of the fold, and it might be several times that big. The quantity of 3.5 per cent ore might be anywhere from 20,000,000 to 100,000,000 tons. The Roan was destined to become one of the great copper mines of the world.

Field and Selkirk sent enthusiastic cables to New York and London. Their clients were at once convinced of the great value of the prospect. They formed the Roan Antelope Copper Mines Ltd. and started to pour in millions of pounds sterling for development and equipment.

With the start of the Roan, the Northern Rhodesia copper boom was on. Selection Trust, Bwana M'Kubwa, and Rhodesian Congo Border Concessions as well as the Roan Antelope soon had dozens of diamond drills at work developing great bedded ore bodies. Parker and the geologists who followed him found that in many places the copper was in the same series of shale and sandstone, always bent into canoe-shaped folds.

By the end of 1929, only four years after Parker first found the chalcocite ore at the Roan Antelope, the six great Rhodesian mines had developed four hundred million tons of ore that averaged more than 4 per cent copper. Later development greatly increased the estimates. Save only for Chile Copper, the Rhodesian mines contained by far the most valuable reserves of copper ore in the world.

The Rhodesian discoveries came just at the time when the copper industry had outgrown the excess in productive capacity that followed the war. Every year consumption made new records. The price of copper climbed to eighteen cents a pound. As usual, copper experts prophesied that it could never drop again. The rosy glow of prosperity made Rhodesian financing easy. The six new mines planned to produce half a billion pounds of copper a year by 1935 and double that amount before 1940. Rhodesia was to be the greatest copper district the world had seen.

Beatty and Davis and their associates in London raised tens of millions of dollars to pay for the development and equipment. English capitalists kept control, in order to break the forty-year domination of the United States over the copper industry, but American engineers bore the brunt of the actual work in Africa. With unlimited funds to draw on, they broke all records for rapid construction.

For a time it seemed that a limit would be set to the expansion by the supply of native labor. The agents of the copper companies scoured the villages to induce the blacks to leave their fields for the crowded compounds and the unaccustomed labor at the mines. At first the new workmen were hopelessly inefficient. For hundreds of years the native population had been undernourished. They did not have strength enough to put in a real day's work even if they had wanted to. And they did not greatly value the things they could buy for the few cents a day the white men paid them in addition to their rations. At the height of the construction period fifty thousand black laborers were doing the work of five thousand American workmen. It seemed doubtful if there were enough natives in all South Africa to carry out the plans. The managers grew old and worn trying to overcome the languid ignorance of the human tools on which they must rely.

As the strengthening food supplied by the companies took effect, the quality of the work improved to a remarkable degree. The natives found they could save enough to buy new wives, and for them wives took the place of a bank account. They learned that a few squares of cloth scattered over the ample black forms of their wives and daughters gave them a social position much higher than that of their naked fellow villagers. It was worth while to work a little harder in order to earn the gaudy things the white men sold. Many natives even became adept at running machine drills in the mines, or the simpler pieces of apparatus in the mills. While the labor at first was expensive at fifty cents a day, after a couple of years it became remarkably cheap. The native did the work of a third of a man instead of a tenth. The peak of the demand for blacks came near the beginning of the construction. Before the plants were completed, the danger of a labor shortage had vanished.

Just as the equipment programs were well under way, along came the 1930 drop in the price of copper and the Depression. Construction was only half financed. It became impossible to raise the rest of the money, as it was evident that the impoverished world could not use the new copper. To avoid bankruptcy, several companies consolidated and spent the resources of all of them to equip one. The tide of English and American engineers flowed back home again.

In spite of the lack of money, the great flotation mills at Roan Antelope, Mufulira, and Nkana started up in 1931 and 1932. Mufulira soon shut down again, and for economy's sake let Roan Antelope take its share in the prorated production. The half-billion pounds' productive capacity that was planned had shrunk to little more than half that amount. Still the Rhodesian properties, thanks to the high grade of ore as well as to the remarkably fine engineering, were making copper as cheaply as any other mines in the world.

HIGH-GRADE ORE IN NORTH AMERICA

The Rhodesian mines alone were big enough to postpone any fear of a copper shortage for several decades. For good measure, Fate came back to America and gave the copper industry four dramatically rich ore bodies of a totally different type from those in Rhodesia. Strangely enough, all four ore bodies were entirely unexpected. Science stumbled upon them accidentally, in places where it anticipated only lean ore. All four were developed within the three years from 1925 to 1928. And all four had so many points of resemblance that these deep high-grade primary ore bodies form the last of the fashions in copper mines.

The Frood ore body of the International Nickel Company near Sudbury, Ontario, was the greatest of these deep ore bodies. For many years International Nickel had gone peacefully along, supplying 80 per cent of the world's nickel from its older mines in the Sudbury District. As a by-product it made two or three million pounds a month of copper. Copper added to the dividends, but International Nickel was not a serious factor in the copper industry.

The greater part of the production came from the Creighton Mine. Year by year the depth of the workings increased until the shaft was more than three thousand feet deep. The ore remained just about the same in size and grade as it had been near the surface. It would apparently continue for several thousand feet deeper. Still three thousand feet is a great depth, and International Nickel decided it must have another mine in reserve to take the brunt of nickel production in case anything happened to the Creighton.

A few miles away the company owned a prospect called the Frood. At the beginning of this century, shallow development had indicated that the Frood contained the greatest body of copper-nickel ore in the world. Unfortunately, the metal content per ton was only about half that in the Creighton. As long as the Creighton could supply all the nickel the world wanted it was not worth-while to spend any more money on its poor relation.

When the postwar boom doubled the consumption of nickel every few years, the Frood was the natural place to turn for more ore. At least it seemed worth-while to put down a few deep diamond drill holes to give a more accurate idea of the tonnage and grade. The drill holes went down to five hundred, a thousand, finally twelve hundred feet depth. There was a slight increase in copper content, but the ore continued to be lean compared with that at the Creighton. A hole was drilled still deeper. To the great surprise of everyone, it found ore that assayed 6 per cent copper instead of 2 per cent. Other holes followed, and the deeper they were the richer the ore became. Shafts and drifts blocked out the ore found in the drill holes. On the twenty-eight-hundred-foot level the great lens, one hundred forty feet wide and many hundred feet long, assayed 12 per cent copper and 2 per cent nickel. In a few months International Nickel developed an ore body many times as great and as rich as the Creighton. The stock market went crazy and boosted the selling price of International Nickel stock to a billion dollars. A few years of hectic construction gave the new ore body a productive capacity of a quarter of a billion pounds of copper a year. The deep, high-grade ore that no one had dreamed of gave International Nickel the greatest copper mine, save only for Utah, in North America. Blind Chance was the engineer in charge.

Luck was just as kind to the Greene Cananea. In the twenty-five years since Bill Greene brought the manners of Tombstone to Wall Street, the rich ore that Jim Kirk had developed in the Capote and the Oversight and the Veta Grande was all mined out. Cananea was struggling along on 2 per cent ore, and there was not much of that. The end seemed only a few years away.

Dr. Ricketts saw just one chance to keep Cananea alive. La Colorada mountain, back of the big mines, received its name from the brilliant red color of the rock from which it was carved. A few prospect holes indicated that the stain was due to the oxidation of disseminated iron sulphide, with which occurred a little copper sulphide. The mineralized material assayed only 1 per cent copper. If the whole mountain assayed as much as this, the lean material might have some value.

To prove whether or not the Colorada body was ore, the Cananea company let a contract to Al Haney to put down a few churn drill holes. One after another the holes developed low-grade porphyry, of doubtful value. As such ore would be worthless at great depth, orders were given to stop all holes at a thousand feet even if they were still in the mineralized mass.

One afternoon as Haney made his last round of the drills he found one of the holes just a thousand feet deep. It was time for the mine superintendent to order the hole stopped. On his way down the hill Haney dropped in at the mine office. It was after office hours, but the boss might still be there. He found that Catron had left a few minutes before to go to a movie. A few feet more or less would not make any difference, Haney thought. He could get Catron's order to stop the hole the next morning.

On the night shift the hole broke into 35 per cent copper ore. There was no more talk of stopping it then. For five hundred feet the drill stayed in high-grade ore. In a few months La Colorada ore body had developed a billion pounds of copper in the richest ore Cananea had ever known. Bill Catron's movie brought Greene Cananea back from the door of the poorhouse to the seat of honor among the scions of Anaconda.

The next prize in the lottery of unexpected ore bodies fell to Calumet and Arizona. Far to the east of the developed mines where the limestone that contains the Bisbee ore is covered by hundreds of feet of barren conglomerate, C. and A. had sunk a prospect shaft called the Campbell. With his usual optimism, Tom Cole insisted that the Campbell must be the biggest shaft in the district, with ore bins and head frame equal to the greatest production Bisbee had seen. For many years it was wasted money. The prospect drifts between the old mines and the Campbell cut only lean sulphide. East of the Campbell, where the engineers knew there was a zone of fracturing that should contain ore, a large flow of water stopped development. No one cared much, as the Bisbee ore had been getting progressively leaner toward the southeast until it was little above the limit of profitable mining.

In a few years, the water was gradually drained out of the porous rock. The eighteen hundred level crosscut east of the Campbell started to creep ahead again. It found the lean mass of iron sulphide the engineers had expected. Smaller bodies within the mass could barely be mined at a profit.

Suddenly a crosscut on the edge of the low-grade area ran into 40 per cent copper glance. For over a hundred feet it continued in glossy black sulphide. The spectacular ore proved to be six hundred feet long and a hundred feet wide, extending from the 1,400 to the 2,350 level. Where only lean material was hoped for, the Campbell ore body turned out to be the richest find in the history of Bisbee. Ten acres of ground at the Campbell would produce as much copper as Calumet and Arizona won from a thousand acres in all its twenty-eight years of life.

Fate had one more surprise in store to confound the geologists. In 1920, Ed Horne and Ed Miller went on a prospecting trip in northern Quebec. In the midst of the forests and lakes and swamps they found some outcrops of gold and copper ore. Noranda Mines, Ltd., took over the claims. The ore bodies were small and not very rich. Still they were close enough to the railroad so that development and equipment was not excessively costly. Noranda built a little smelter and started to cash in on the high copper price of 1928.

The first shafts were too close to the ore for comfort. To be safe, Noranda started a main working shaft halfway between the two principal ore bodies. As the only mineralization near the shaft was a very lean outcrop of silicious pyrite, there was no danger of finding ore close enough to interfere with the use of the shaft for permanent hoisting.

Development had shown that most of the small ore lenses played out at four hundred feet depth. There was a chance that ore might come in again, and the Noranda directors decided to sink the main shaft to a thousand feet. If there was no more ore in depth, the property was of very little value anyhow, and the money that would be spent in sinking would hardly be worth distributing to the stockholders.

At nine hundred and sixty feet the shaft entered high-grade copper and gold ore. A crosscut a few feet deeper found one hundred and twenty feet of 14 per cent ore, carrying nearly $10.00 a ton in gold in addition to the copper. To the complete surprise of all the experts the mineralization with lean, worthless pyrite had changed in depth into a great bonanza. The "H" ore body was worth more than all the rest of the Noranda property.

Encouraged by the four deep high-grade ore bodies of International Nickel, Greene Cananea, Calumet and Arizona and Noranda, geologists worked out fine theories to prove why the ore was there. They even ventured to prophesy a lot of other similar lenses. The reasoning was so very persuasive that the stock market discounted two or three more ore bodies like La Colorada and the Campbell. But the pattern was lost. The millions spent in development have not found a single new great ore body of the deep, high-grade type.

The wave of optimism that swept the copper world from 1925 to 1929 carried even Fate off her feet. We needed new copper mines, but not so many new copper mines. The simultaneous discovery of the Rhodesian deposits, the deep high-grade lenses, and the big north extension of Chile Copper was too prodigal a blessing, like a desert cloudburst after a year of drought.

The ore developed in four years contained as much copper as the world could use in twenty years, with consumption at the 1929 rate. Even if the demand continued to increase 10 per cent a year, as the optimists hoped, the new copper would have lasted ten years. The God of Prospectors must have become so interested in his new styles of ore bodies that he did not notice until too late that he was running off too many copies of them.

The copper magnates did an ever poorer job of guessing. They ought to have known that consumption could not increase faster than 10 per cent a year at best. If they had been really far sighted, they could have anticipated a slowing down of the rate of increase, or even a moderate-sized depression. But the high-grade ore was an irresistible temptation. It led the operators into a wild orgy of optimism. They equipped the new mines so extravagantly that they were soon ready to produce half again as much copper as the world had ever needed. Even with a normal growth of consumption their bad judgment must have brought hard times to the industry.

THE GREAT DEPRESSION

On top of the poor judgment of the copper men came the great Depression. Instead of increasing 5 or 10 per cent a year, the world's demand for copper decreased 25 per cent a year. The new mills and smelters started up just when their added supply of metal would do the most harm. Before the producers realized what was happening, they had accumulated a year's supply of refined copper. The good old law of supply and demand was waiting around the corner with a big club. Five-cent copper was the result.

This meant starvation for the copper mines. Most of them couldn't approach a five-cent cost of production. Many mines shut down in despair. Others struggled along at a small percentage of capacity to keep their employees from going hungry. They were wasting their ore reserves and depleting their treasuries in the vain hope that they could keep their heads above water until some miracle doubled the selling price.

Only five or six copper mines in the world could make copper for the five cents a pound for which they sold it for nearly a year. Even they had to resort to skillful manipulation of the books to avoid showing a loss. Chile, Braden, Utah and the two operating Rhodesian mines were not actually out of pocket for the copper they sold. The Rhodesian properties, helped by the low exchange, claimed to be making copper for three and a half cents. But even these remarkably low-cost mines were using up their best ore and their millions of dollars' worth of new machinery without recompense. In a few years they would have been compelled to mine leaner or less accessible ore and to replace worn-out equipment. Then if five-cent copper continued they would have joined the higher cost producers in a universal bankruptcy.

As the lowest depths of the Depression passed, the copper situation became a little less desperate. Aided by agreements to curtail production, the metal price climbed to eight or nine cents a pound in the United States. There was a smaller increase in the foreign price. With the improvement in world business, the overwhelming surplus of unsold copper had decreased a little. Most of the great companies were not actually losing money from day to day, even if they were throwing away their ore. If they could hold on without bankruptcy, they could at least hope that the future might have better days in store.

Sad as was the present state of copper mining, in the past it had recovered from even more hopeless despondency. Other periods of overproduction and low price have in the long run helped copper far more than they hurt it. In the lean years from 1880 to 1893, Butte and the Arizona mines added many times as much copper to the world's reserves, in proportion to the small consumption, as the discoveries since 1925 have given us. The resulting ten years of low prices stimulated consumption so much that they made copper mining a great industry. Again Jackling's success with Utah brought about the almost immediate development of enough copper to last twenty years, at the highest rate of consumption known up to that time. Only the World War prevented a period of overproduction like that of the 'eighties. We know now that even without the war a few years' normal growth of industry would have caught up with the great new reserves and productive capacity.

The Bonanza Mine of Kennecott, Cerro de Pasco, and Katanga in turn threatened to wreck the copper industry. Almost before they reached full production their rich ore proved a blessing to an age that was hungry for electrical machinery instead of a curse to the other copper mines.

Now the world seems once more completely swamped with copper. But the same growth of industry that caught up with the still more menacing discoveries twenty years ago and forty years ago will come to the rescue again. The curve of mechanical progress may be flattening, as many think. But unless our whole industrial civilization ends in a gigantic catastrophe, we shall continue to use more and more metal. The Depression years are storing up an unsatisfied demand for copper as for other commodities. When the gates of prosperity once more swing open as they have so often done in the past, the unsold stocks of copper that seem so great will vanish before the miners realize what is happening. Stimulated by the low price, the uses of copper will expand far more than ever before. The mines that are ready to take advantage of the demand, with costs lowered by years of adversity, will reap profits higher than they can dream of.

Some will of course fall by the wayside before prosperity comes again. The great enriched desert ore bodies will soon be completely exhausted. Butte is no longer the Richest Hill on Earth. Many of the deep mines of the Michigan Copper Country have hoisted their last skipload of ore. Even the smaller porphyry coppers have reached the point where decreasing grade has brought the end of the trail in plain sight.

The greatest disseminated deposits – Utah, New Cornelia, Chile and the rest – will be ready to join Rhodesia in celebrating the return of the reign of King Copper. The ore reserves that now seem so hopelessly great will soon melt away. Before we realize it, the world will again be dreading a copper famine. Engineers and geologists will once more comb the far ends of the earth for a new Roan Antelope or another Chile Copper. And once more the Fate that watches over prospectors will choose the unexpected path and will add a new fashion in ore bodies to the romantic story of the copper mines.

AUTHOR'S NOTE: While recovery of the copper industry from the great Depression took longer than I expected when I wrote *Romantic Copper*, the recovery did come and was much more dramatic than anyone had foretold. The forty years since 1933 have seen a greater growth in copper use and production than all of the preceding centuries. *Romantic Copper* really just set the stage for the events described in the second part of this book.

PART II

Foreword

THE romantic part of the story of copper closed with the prophecy that a new threat of a copper shortage would follow the overproduction of the Depression years. There would be a new search for ore bodies, and the search would result in New Fashions in Mines.

The prophecy proved to be true. The intensive search for more ore in the old mines and for new types of mines found ore bodies that differed greatly from those of earlier years or centuries. Both the number and the size of the new discoveries were far greater than anyone expected. As much copper has been found since 1935 as all that was mined in the preceding thousands of years. And growing consumption in our mechanized society has kept pace with the output of the new mines. While there may be occasional overproduction, the great problem in the future will be to keep on finding enough copper.

By far the greatest additions to our copper supply in the past generations have come from enormous deposits that are so lean that no one thought of them as ore before 1930.

Dr. L. D. Ricketts had paved the way for the great but extremely low-grade copper ore bodies. Fifty years ago he realized that the cost of making copper was made up of "ton costs" and "pound costs." "Ton costs" include mining, milling and other expenses that depend on the tonnage mined. "Pound costs" include smelting, refining and most overhead expenses. Both of these are the same regardless of how rich or lean the ore may be. In determining whether or not material is ore, the first step is to subtract the expected "pound costs" from the probable selling price of the metal. The balance is available for "ton costs" and profit.

For instance, with the present value of the dollar reduced by inflation, it is reasonable to assume a future copper price of 60 cents per pound. "Pound costs" will seldom exceed 15 cents. This means that 45 cents per pound are available for "ton costs" and profit. If the deposit is big enough to stand operation at the rate of 10,000 or more tons of ore per day, "ton costs" should not exceed $2.00. They may be much lower. This means that a recovery of less than four pounds per ton will break even.

This reasoning, that seems so simple, changed many billion tons of worthless rock into ore. While those who started the new low-grade mines have seldom thought as clearly as did Dr. Ricketts, they reached the same conclusion. And all of us enjoy the electricity and other amenities that ample copper brings.

The flood of new copper in the past thirty-five years has come more from thinking and technological progress than from romance and adventure. This situation almost bears out the conviction of my promoter friend, Johnny Baggs. Johnny was sure of the power of mind over matter: if by "constructive thinking" he followed in his mind all of the processes of ore formation, the ore would be there. I did not know of this belief of Johnny's when I started to develop an abandoned silver mine at Austin, Nevada, forty-five years ago. I carefully mapped the geology and found a place where there should be bonanza ore like that which had made Treasure Hill famous. Johnny had come to the same conclusion by "constructive thinking." I ran a tunnel to the hoped-for ore body and found – an old stope that had caved before it was recorded on any maps. My geology was correct, but the ore was gone. Johnny maintained that all "constructive thinking" could be expected to do was to find where the ore should be. If only a hole in the ground was left, this wasn't the fault of his control of mind over matter.

So the reasoning that makes five- or six-pound material into ore is correct, unless there is some unexpected factor such as an ore from which the copper cannot be recovered or a prohibitive flow of water or any of the other ills that sometimes wreck the best hopes. Fortunately such troubles have occurred far apart and have not prevented Dr. Ricketts' reasoning about "ton costs" from doubling the world's supply of copper.

In the chapters that follow I have tried to give credit to the vision and venturesome spirit that started the new types of ore bodies. The ones who came after didn't have to have as much imagination. They found most of the ore but they didn't have as much fun as the ones who started the New Fashions in Mines.

9

What Happened to the Old Copper Mines?

When it became evident, about 1935, that copper consumption was catching up with productive capacity, the easiest way to get more copper was from the great old mines. As the prospector says, the best place to find copper is where copper is, and the old mines have turned out unbelievable amounts of metal. Most of the increased ore was due to the application of Dr. Ricketts' idea of ton costs; more efficient mining and ore treatment helped to change barren material into ore. The improvements entailed hundreds of millions of dollars spent for new plants.

The change took place all over the world. In the United States, Kennecott, Phelps Dodge and a handful of others have increased their rate of production by 60 per cent since 1929; the average grade of ore mined has been cut in half. Large Latin American mines increased production from 1929 to 1969 by an average of 52 per cent; the proportionate fall in grade was as great as that in the United States. Annual output of the older mines in the Katanga deposits of the Congo increased by 150 per cent in the forty years. The four great mines in Northern Rhodesia – now Zambia – hardly had a good start before the Depression forced them to curtail, but in 1969 they turned out nearly 700,000 tons of copper. This was 11 per cent of the entire world production. Without these rapid increases in the older mines the world could not possibly have met the requirements of World War II, as well as the increasing consumption of copper for peaceful expansion. The older mines saved the day.

It is remarkable that out of about thirty large-scale copper producers in 1929, only four have been exhausted. And three of these may come back.

Calumet and Hecla is the greatest of those that have failed. This wonderful old mine struggled along until 1968, making a little money when the price of the metal was high, and losing during depressions in the copper market. A costly attempt to reopen the old Calumet and Hecla Conglomerate with government help failed; the cost of pumping the millions of tons of water that filled the old workings from the 9,600 to the 3,100 foot inclined level was too great.

Before the shutdown a new ore body was developed northeast of the old C. and H. Conglomerate. Above the 3,000 foot level this Centennial ore body had been too small to pay the cost of development, but it seemed worth opening up to greater depth. On the 3,600 level it expanded to over a two thousand foot length of profitable ore. Drilling deep holes from the surface and extending old exploration drifts from the old mine made it seem likely that many million additional tons of ore remain in the Conglomerate bed.

The future looked bright until a labor dispute upset the apple cart. The cost of living — and wages — had always been lower in the Copper Country of Michigan than in mines further west, but the Michigan Miners' Union wanted the western copper scale. They shut the mine down by a strike. The miners preferred to go on relief or to drive more than a hundred miles a day to and from the White Pine mine rather than to take subwestern wages at Calumet. After years of fruitless negotiations Universal Oil Products Company, which had bought Calumet and Hecla, gave up the fight, shut down the pumps and sold much of the equipment. So the hundred-year story of Calumet and Hecla ended, at least for the present. There may be a billion pounds of recoverable copper in the Calumet Conglomerate, but the mine was worthless under present conditions. Only a completely new technology could bring back the Calumet and Hecla, most engineers thought.

There were still some optimists left. The Centennial workings, added to many miles of drifts along the Conglomerate Lode northeast of the old workings that had been run many years ago, indicated tens of millions of tons of ore that might average well over one per cent copper. This was twice as much as the content of the ore that was being mined in all the great open-pit deposits. Ira Joralemon, probably the only engineer left alive who had seen the bottom of the Calumet and Hecla, had made an estimate of the ore that might be hoped for. After long consideration, the Homestake Mining Company made a tentative contract with Universal Oil late in 1972 that would allow them to operate the Calumet mines. In addition to the continuation of the Centennial workings to greater depth, Homestake thinks there is a chance that a new process for dissolving the copper in the Conglomerate Lode by solutions introduced through drill holes may avoid the necessity of deep new shafts. If, as seems likely, the Calumet miners are tired of living in a "ghost town," the Calumet and Hecla Conglomerate mines may once more be among the great copper producers.

The Old Dominion Mine in Globe, Arizona, seems just as dead. In the bottom of the mine, below the 4,000 foot level, the ore had become lean and costs were prohibitive. The only hope was at the west end of the mine, where the great Pinal Fault had cut off the good ore, but attempts to find a faulted section beyond the fault were frustrated by great flows of hot water. The Depression of the 1930s put the finishing touches on the fifty-year-old mine. The only hope is that mining a deep faulted extension of the great Miami-Inspiration ore may drain the ground in which the rich Old Dominion vein may continue.

The Pilares mine of Phelps Dodge, near Nacozari in Sonora, Mexico, was comparatively small, never producing much more than 20,000 tons of copper a year; its profitable life was about 25 years. Soon after 1929 the better ore was exhausted, and the mine and costly plant were abandoned. This is the most hopeless of the worked-out copper mines. But to make up for it, several hundred million tons of open-pit ore have recently been found a few miles to the southeast of Pilares. While the original mine is finished, the district has just begun.

The United Verde and United Verde Extension mines at Jerome, Arizona, complete the list of great worked-out copper mines. These two are really faulted sections of the same nearly vertical group of lenses of rich ore. The ore in each mine is within an area less than fifteen hundred feet in diameter. Allowing for the fault, the original vertical extent was nearly six thousand feet. The United Verde Extension fault block is, as far as is known, completely worked out. "Leasers" still make a precarious living gouging out remnants of ore from the walls of the great United Verde open pit. The handful of old-timers and the "hippie" squatters who still live in the decaying houses of Jerome call it the "Greatest Ghost City in the World."

It is hard to believe that there is no ore at Jerome outside of the fifty acres that held the two great mines before the fault cut the ore in two. In the present inflated currency, the profit from the two mines amounted to nearly half a billion dollars. The United Verde Extension and Phelps Dodge, which had bought the United Verde, ran many miles of drifts and drill holes trying to find more ore, but they both failed. The only result was a large body of copper, zinc and iron-bearing material in lower levels of the United Verde, and this was too lean to justify the costly plant that it would require. After much thought Phelps Dodge abandoned it.

In the early days of the "UVX," half a dozen stock companies sank deep shafts on all sides of the old mines and spent more than ten million dollars in fruitless searches for a new bonanza. In the past twenty years the successors to UVX have given options to several of the great mining companies of the United States and Canada. Guided by the newest geophysical and geochemical tools and by new theories devised by their capable geologists, they have spent more millions. And all have failed. Other companies will certainly try. The possible prize is too tempting to be abandoned permanently. New theories will seem convincing even though "only a handful are still alive" who saw the old ore bodies in the mines that are now caved or full of water. Somewhere, under the many square miles of rugged mountainsides, covered by rocks that were laid down after the ore formation, there must be another UVX. Some day a new method of probing the earth's secrets may solve the enigma. Until that day comes, Jerome will continue to be a mausoleum of futile hopes.

This photograph shows the Old Dominion Mine and smelter at Globe, Arizona, in 1902. It eventually became a Phelps Dodge property. (—MINING AND SCIENTIFIC PRESS)

Making an electromagnetic survey for new copper-nickel deposits, a Twin Otter tows a sixty-pound fiberglass magnetometer "bomb." Below, the plane's interior shows cable, reel and instruments. *(—All photographs in this chapter, International Nickel Co. of Canada, Ltd.)*

10

First of the New Fashions—Lenses of Ore in Eastern Canada

FOUR dead copper mines out of the thirty great producers in 1929 is a pretty good record. With this past success it is no wonder that the companies started an almost frantic hunt for new deposits soon after the end of the Depression of the 1930s. And as in earlier periods of threatened copper shortage, the hunt has yielded spectacular results: nearly eighty great new copper mines, scattered all over the world, have been found since 1935. As in earlier periods, also, the new mines came in groups. Several deposits of similar types, as though following New Fashions in Mines, have been developed at almost the same time.

The search for new sorts of copper deposits was stimulated by an increase in price of the metal. The price soared from five and one half cents per pound in 1932 to nearly fifty-eight cents in 1970. In 1969 the London price reached the equivalent of more than seventy cents per pound, and soared for a brief time to nearly a dollar a pound. It was no wonder that world production of copper increased from a million tons in the depression year of 1932 to 6.9 million tons in 1969. The boom in copper led the companies to spend millions of dollars hunting for new ore bodies. And they found them.

The first of the new fashions in Copper Mines included lenses of high-grade ore, chiefly in eastern Canada. Noranda had shown the way before the Depression; most of the new mines contained valuable other metals as by-products.

If there was any real hero in the story of the Canadian ore lenses, it was the Canadian tax law, which allowed new mines

On the ground, the plane stows the "bomb" in its rack. Below, maps are prepared from the survey data.

to produce for three years before income taxes began. Money spent for development and equipment could be returned, at least in large part, in the three years. Companies and individuals could afford to take longer chances in their exploration; both the older companies and new venture groups scoured the country with all the new geophysical aids. The income tax law has been a spectacular success, and future taxes will far more than make up for the deferment. Unfortunately, the legislators have been impatient; they have recently done away with the three-year freedom from taxes on new mines. They will pay for it by fewer new mines.

The nickel-producing companies, International Nickel and Falconbridge, led the way. A threatened acute shortage of nickel stimulated the search. The companies developed the art of finding ore with delicate magnetometers suspended below small airplanes. These "airborne magnetometers" could detect indications of ore even under a thick cover of glacial gravel. The two nickel companies have found great new nickel-copper ore bodies, both in the old Sudbury district of Ontario and further north and west, near Hudson Bay. As a result, instead of an acute shortage, there is an excess productive capacity of nickel, and the price of shares in the companies has plummeted since the end of 1970. Sherritt-Gordon and the Hudson Bay companies found other ore bodies that contain much copper, although nickel and other metals are often more important. These discoveries have been due to the painstaking work of many engineers and geologists; they have not been as exciting as discoveries due to one or two men with vision, but the ore is just as valuable.

Except for the copper-nickel ore bodies, the first few years of the Canadian wave of exploration found ore bodies of moderate size. Opemisca, Campbell-Chibougamau and a few others began to produce copper successfully in the mid-1950s, although they were not big enough to make a dent in the general copper situation. Then larger deposits were found by the same careful study. Gaspe Copper and the Geco division of

When the maps are put t
gether to form an electroma
netic "picture" of the are
the giant drill rig at left go
in to inspect anomalies.
drills unusually deep holes.

Noranda started on a small scale before 1960 and built up their production to more than 32,000 tons of copper per year each.

Then came the jackpot. In addition to the lenses of copper ore, exploration in eastern Canada had found large deposits of lead and zinc ore. The greatest were in the Mattagami Lake mine north of Noranda, owned largely by Noranda and Placer Development. This great ore body stimulated the interest of other mining companies. They crisscrossed a great area of wilderness covered by forest and glacial detritus with magnetometer surveys. The Texas Gulf Sulphur Company found several magnetic anomalies northwest of Noranda – an anomaly being "something different." To check the anomalies they drilled costly diamond drill holes through the detritus, but for several years the search was fruitless, and Texas Gulf's associates became discouraged and dropped out. Then, as hope was waning, another drill hole broke into fantastically rich ore. It cut many hundred feet of material that contained so much copper, zinc and silver that any one of the three would have made it a rich mine.

The rest was easy, as far as ore was concerned. Drilling blocked out one of the great ore bodies of all time; much of it could be mined by open pit. Texas Gulf built a 9,000 ton per day mill at this "Ecstall" mine in record time. The difficult problem of separating the different metals was solved, and in 1967, the first full year of operation, the Ecstall Mine produced 49,200 tons of copper, 224,640 tons of zinc, and 13,968,000 ounces of silver. No other mine in the world produced as much silver or zinc. Production has fallen off a little from the peak, due to a decreasing grade of ore and a deeper pit. By the end of 1972 the limit of ore that can be economically mined by open pit in the Ecstall Mine was not far ahead. The company is gradually changing to underground methods. The profit per ton may not be quite as great, but the spectacularly rich ore will for many years continue to make the Ecstall one of the greatest of all the "base metal" mines.

Unfortunately there is always a price to pay for a great gain, and in the case of Texas Gulf the price was litigation.

A modern L.H.D. loading and tramming machine replaces shovels and cars in an International Nickel Company mine.

Leitch Gold Mines, a comparatively small Canadian company, had participated in early stages of the Ecstall exploration, and after the great ore body was found, Leitch brought suit on the grounds that its interest had not been formally cancelled. The Ontario Supreme Court decided the case in favor of Texas Gulf in November, 1968.

An even more serious suit was brought by some Texas Gulf stockholders and the United States Securities and Exchange Commission. They claimed that directors, officers and the geologist who was largely responsible for finding the Ecstall had bought Texas Gulf stock after the first drill hole had cut the ore, but before stockholders had been notified. Eminent American and Canadian engineers testified in the long suit. The question was: How much ore can be estimated from one good drill hole after many holes had failed to find ore? This is obviously a matter of opinion, not of fact. Regardless of the outcome, Texas Gulf profited enormously from the foresight and persistence of its officers and geologist. These employees took a chance in buying the stock, as the first drill hole might have chanced to hit a small, rich seam of ore.

This Texas Gulf suit has hurt the mining industry, upsetting the custom of many years. The fortunes of Herbert Hoover, D. C. Jackling, Chester Beatty and many other giants of the mining world were based on purchase of stock in their companies at favorable times. The successes were well advertised and induced many of the brightest young college men to go into mining. However, the spectacular fortunes of the few lucky ones were insignificant compared with the wealth the discoveries brought to the stockholders. Today a responsible employee or officer hardly dares buy stock in his company at any time. If the stock goes up, he can be accused of buying due to confidential favorable information, or if it goes down, other stockholders can say he sold because he knew of results that were not rosy. In either case, losing a possible profit or failing to avoid a loss might cause a stockholder to sue. And there is no way to prove just when development has proven that there will be either success or failure.

The Texas Gulf suits have already prevented good men from accepting positions with mining companies. Salaries alone are not high enough to attract the most capable engineers. If the best men cannot profit from the great ore bodies they help develop, mining is not worth-while. So mining and exploration no longer get the men with the most vision and energy. Indeed, misrepresentation or deliberate withholding of information should be punished, but the willingness to join a venture with the company by buying stock, when success is not certain, should be encouraged. We need the best men, and the chance of great financial success is the most certain inducement.

With the Ecstall, lenses of high-grade copper ore in Canada went out of fashion, but another type of ore body took over just in time to avoid a copper shortage.

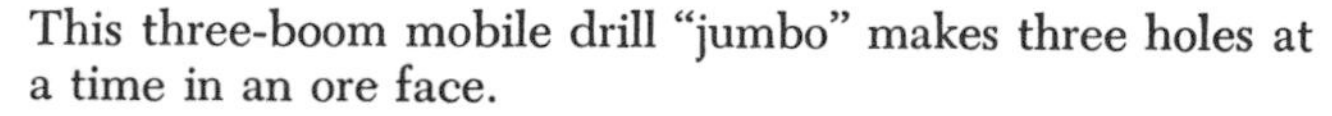
This three-boom mobile drill "jumbo" makes three holes at a time in an ore face.

11

Great Low-Grade Copper Deposits in the United States

THE new "fashion" included many great but exceedingly low-grade copper deposits. In the twenty years following 1950, dozens of these "disseminated" copper deposits have been found in many parts of the world. The ore bodies generally contain from tens to hundreds of millions of tons of ore that averages from 0.5 per cent to 0.8 per cent copper, and most of this ore can be mined cheaply by open pit.

These ore bodies would have been worthless before the advent of the new large-scale mining equipment and before concentration by flotation was perfected. Power shovels that take up tens of tons at a bite and trucks that carry loads of more than a hundred tons have allowed a decrease in open-pit mining costs to less than twenty-five cents per ton, and by flotation more than 90 per cent of the copper in lean ore is recovered. Ton costs dropped unbelievably. One per cent ore is now more valuable than 2 per cent ore was thirty years ago.

Two widely separated ventures took advantage of Dr. Ricketts' analysis of costs. The first, in Arizona, was the result of detailed technical study, with just enough adventure thrown in to make it interesting. The second, in British Columbia, came from the vision and persistence of an old-fashioned prospector.

The widespread operations and large staffs of the great mining companies make it hard to pin the credit on one or two men. But Louis S. Cates, president, and Harry Lavender,

The shift is about to go down in the Cole Shaft, one of four underground mines still active in Bisbee after more than ninety years. The Lavender pit in 1971, below, shows Bisbee at top, concentrator at lower right. *(—Both, Copper Queen Branch, Phelps Dodge Corp.)*

manager of operations of Phelps Dodge thirty-five years ago, were probably the ones most responsible for the development of great low-grade ore bodies in the United States. They combined engineering skill with an intuition of what risks to take, and when.

BISBEE, MORENCI AND TYRONE

In the 1930s the high-grade ore that had made Morenci and Bisbee among the greatest of copper districts was approaching an end. But in Morenci there was an enormous tonnage of very low-grade material, and in Bisbee a great but smaller tonnage. The grade at both places seemed too low to justify production on the large scale that would be needed to achieve low costs. Lavender found that his New Cornelia plant at Ajo, Arizona, was just the right place to prove the value of lean material. Under 1935 conditions, with ten-cent copper, he made a profit at Ajo out of ore that contained only seven pounds of copper per ton. This meant that the Clay body at Morenci and the Sacramento Hill area at Bisbee were ore. Cates and Lavender built a 20,000 ton per day plant at Morenci and a 12,000 ton plant at Bisbee. The Clay ore body was in full operation in the mid-1940s and the Lavender pit in Bisbee in 1954.

The success of these costly ventures proved that material that carried 0.75 per cent copper was good ore. The Lavender pit has limited reserves and will soon be exhausted, but the Clay ore body has hundreds of millions of tons of ore. It will last for decades with the present production of nineteen million tons of ore per year.

The Hanna Mining Company, with its exploration then run by Mack Lake, shared in the Morenci success. It acquired a group of claims on the almost precipitous slope of Markeen Mountain, across Chase Creek from the Clay ore body. Drilling found a hundred million tons of lean ore, and Phelps Dodge developed a continuation of the ore in their claims beyond Chase Creek. This included the old Metcalf and King un-

Above are the concentrator and smelter of Phelps Dodge at Morenci; below, the Clay ore body. The ore trains there, so small they can hardly be seen, indicate the enormous size of the open pit. *(—Both, Morenci Branch, Phelps Dodge Corp.)*

derground mines, run by the Arizona Copper Company many years earlier. Hanna thought its ore body a little too lean to justify spending the hundred million dollars it would take to bring it to production; after long negotiations Phelps Dodge leased the Hanna ground. It is stripping off the barren "capping" from the mountainside over the ore and is building a great flotation mill. This new Metcalf Mine will be independent of the older Morenci mining operations. Royalties to Hanna will make its Chase Creek venture a great success.

The Clay ore body profits whetted Phelps Dodge's appetite for great low-grade mines. Sixty years ago it had bought the old Burro Mountain Copper Company, south of Silver City, New Mexico. Phelps Dodge had found a limited tonnage of 1 to 2 per cent copper ore near the lenses of richer copper and iron sulphides that had made the Burro Mountain one of the early unsuccessful Southwestern copper camps.

At about the same time, Chester Congdon and others who had shared in the success of Calumet and Arizona bought claims northeast of the Burro Mountain, calling their new town Tyrone. With three to six hundred foot shafts, thousands of feet of drifts and a few drill holes, their Chemung Copper Company developed a couple of million tons of 2 per cent ore. I. J. Stauber and others found similar ore at the adjoining Savannah Copper Company. The lenses of ore were a little too small to yield a profit. In 1912 and 1915 Phelps Dodge – then the Copper Queen – bought the two other properties for less than two million dollars.

The trouble wasn't over. Gene Sawyer, one of the ablest young Phelps Dodge engineers, went to Tyrone as manager. He found a few more million tons of ore that must be mined underground. The company built a 2,000 ton per day mill and a beautiful new town that was designed by a leading New York architect. Sawyer mined two million tons of ore, but the profit was insignificant, largely because the mill could not recover part of the copper that was in oxidized form. The collapse of the copper price in 1921 forced a shutdown. Saw-

An ore train is on its way from the Clay open pit to the Morenci concentrator. The power shovel below loads ore at the Clay. In most open pits trains have been replaced by trucks. *(—Both, Morenci Branch, Phelps Dodge Corp.)*

yer worked out a successful method of leaching low-grade material on dumps that recouped part of the earlier loss. He devised what he thought would be an economic method of recovering much of the oxide as well as the sulphide copper in the 1.5 per cent to 2 per cent ore Company officials were not convinced.

Tyrone remained shut down until long after Sawyer died, but drilling continued and the ore body kept on expanding and getting lower-grade. By 1960 several hundred million tons of 0.75 per cent ore had been developed. Large-scale tests at Morenci and Ajo proved that with new techniques much of the Tyrone oxidized copper could be recovered. With success assured, Phelps Dodge built a 25,000 ton per day mill at Tyrone, in addition to a new town for employees, a twelve-mile railway and all facilities. The mill started in 1969 after $118,000,000 had been spent; everything worked so perfectly that the capacity is being increased to 100,000 tons per day. A smelter is being built to treat the concentrates. Tyrone is finally one of the great copper mines of the world, but it took sixty-five years and more than a hundred million dollars from the time Nathan Leopold failed at the Burro Mountain Copper Company.

It seems at first glance that the sale of the Chemung and Savannah companies to Phelps Dodge for less than two million dollars was a spectacular mistake. Actually, the sellers got a good bargain. At six per cent interest the $1,500,000 they received amounted to $35,000,000 in 1969. Inflation increases this to a hundred million. Each of the early owners owned only part of the ore body, which would have increased the difficulty of mining and the cost of the plant, and for fifty years success would have been uncertain. In 1912 the author, who had been responsible for the development of the New Cornelia Mine, advised Chester Congdon and George Tener to sell to Phelps Dodge. In spite of the hundreds of millions in profits that Phelps Dodge will make out of Tyrone, hindsight shows that the decision was correct.

The outcrop of the new Metcalf Mine, above, is across Chase Creek from the Clay at Morenci. Below, the open pit of Asarco's Mission Mine is shown in 1970. *(—Above, photograph by the author; below, Asarco Annual Report)*

SILVER BELL, PIMA AND MISSION

In the past twenty years more than a dozen other successful low-grade copper deposits have been found in the United States. In Arizona the Silver Bell ore body of the American Smelting and Refining Company had been known before 1950; it has tens rather than hundreds of millions of tons of ore. The low-grade and the oxidized form of part of the copper delayed production, but the company finally got up its nerve and built a 7,000 ton per day mill. Production started in 1954, a few years after the Clay ore body and the Lavender pit. It has been a real but not an overwhelming success.

Pima Mining Company, twenty miles southwest of Tucson, came in three years after Silver Bell, and geophysical exploration was responsible for the discovery. The United Geophysical Company, under the leadership of Herbert Hoover, Jr., had been successful in exploring for petroleum and thought it could use its knowledge in finding ore. It went about the job with intelligence rather than dramatic vision; so first it searched through technical articles and government publications. The Twin Buttes area southwest of Tucson seemed promising; in this old district small copper and lead-zinc silver deposits had been known for decades. United Geophysical thought there might be larger ore bodies under a thick cover of soil or "desert wash" between the Mineral Hill copper prospect of Banner Mining Company, the San Xavier lead-zinc mine of Eagle-Picher, and the old Twin Buttes limestone replacement lenses. They sent Walter Heinrichs to make a magnetometer survey of the desert area. This found a strong anomaly east of the Mineral Hill property line that was worth drilling.

The first hole cut 3 or 4 per cent copper sulphide ore. Further work showed that the sulphides were in altered limestone and shale, cut by porphyry dikes so highly altered as to be hardly recognizable. Drilling soon developed more than a million tons that averaged 4.5 per cent copper. East and south of the rich ore, drill holes found disseminated copper sulphides in impure, limy sediments cut by porphyry dikes. The indicated copper content was little more than 0.5 per cent.

The richer ore was worth the cost of preparing for production on a modest scale, but the job was too big for United Geophysical alone. The Cyprus Mines Company, controlled by the Mudd family of Los Angeles, and the Utah Construction and Mining Company of Salt Lake joined in the venture, with Utah as the operator. Because of the "heavy ground" and the irregularity of the richer ore, Utah decided on open-pit mining. It built a 1,000 ton per day mill and started successful production in 1959.

Meanwhile the low-grade "disseminated" ore body east of the richer ore continued to grow. Successive additions were made to the mill, and capacity is now being increased to 53,000 tons of ore per day. Production will be at the rate of 80,000 tons of copper a year. With two hundred million tons of 0.5 per cent copper ore in reserve, Pima is one of the great copper mines.

The chief difficulties met by Pima were due to contests about property ownership. The richer ore extended into the adjoining Mineral Hill property of Banner Mining Company, and the good ore near the property line could not be mined by either company without trespassing on ground of the other. Banner had no great investment in plant, so could afford to wait, and the final agreement was favorable to it. Further trouble came from a contest as to whether State leases, on which much of the ore occurred, were valid. Pima finally got a clear title. The delay was costly but far from fatal.

A far more serious loss came from the generosity of Pima in letting engineers of other companies visit their property and even see the drill cores. As the value of the low-grade disseminated deposit east of the richer ore became evident, Pima started to locate a large group of claims out in the desert. But American Smelting and Refining Company engineers had seen the Pima drill cores, and with this knowledge they located for Asarco a great area farther northeast. Much of the low-grade ore extended into the Asarco claims, which were called the Mission Mine. Mission built a 20,000 ton per day mill that started full production in 1962; mining is by open pit. The Mission Mine, with a production of 50,000 tons of copper per year, is the most valuable Asarco mine in the United States.

This loss to Pima of one of the great copper mines makes one wonder whether generosity in welcoming visitors is worthwhile. Sixty years ago, due partly to apex lawsuits, all information was carefully guarded. Dr. James Douglas and Dr. Louis D. Ricketts were largely responsible for a more liberal policy, although they realized that now and then someone would take an unfair advantage of the free information. But they were sure that in the long run an interchange of knowledge would benefit everyone, and in most cases they were right. Pima suffered by the policy, but if Pima had not known of the success in mining very low-grade ores at Morenci, it might not have acquired any part of the Twin Buttes District. As in nearly all our endeavors, we must weigh a gain against a possible loss.

After the success of the Mission Mine was assured, Asarco paid more than a million dollars to the Pima Indian tribe for an area adjoining the Mission. Drilling proved that the Mission ore extended into this San Xavier Mine, but it was nearly all oxidized. Asarco is now building a leaching plant to treat this ore. The leaching process will have the additional advantage of using some of the sulphuric acid that must be made from the gases from the Asarco smelter at Hayden, Arizona. The leaching plant will thus reduce pollution by smelter gases.

Two other projects in Arizona will add to the copper production of this great company. The Sacaton Mine near Casa Grande is in a mineralized, steeply dipping fracture zone rather than a typical disseminated deposit. Much of it must be mined underground. A mining and flotation plant with a capacity of

The three who made Duval great are W. P. Morris and George E. Atwood, above left and right respectively, and Harrison Schmitt, shown at left in the field. *(—Above, Duval Corp.; left, Mrs. Harrison Schmitt)*

9000 tons of ore per day is under construction. It is estimated that 46,000,000 tons of 0.75 per cent copper ore will be mined.

The last Asarco project in Arizona is in the desert southeast of Florence. Asarco drilled comparatively small outcrops and found a few tens of millions of tons of ore that averages more than half a per cent copper but is too narrow and deep for open-pit mining. It seems an ideal place for testing a process devised by metallurgists of Asarco and Dow Chemical Co. for leaching through drill holes after shattering the ore by introducing water under enormous pressure. If it works, this process also will reduce atmospheric pollution by sulphurous gases.

DUVAL MINES

Other low-grade copper deposits that came to production a little before and after 1960 radically changed the copper picture in the United States. They emphasized the fact that two or three top men with vision and judgment can make a company.

A few decades ago the United Gas Company of Houston found a sulphur deposit while hunting for natural gas and turned it over to a subsidiary to operate. This company soon found a good potash mine near Carlsbad, New Mexico. The subsidiary was called Duval Sulphur and Potash Company, later Duval Corporation, and United Gas owned more than seventy per cent of the stock.

The president of Duval, W. P. Morris, and the technical chief, George Atwood, wanted more successes. As they knew little about metal mining, they employed Harrison Schmitt, a consulting geologist from southwestern New Mexico, to help them in their new ventures. Morris and Atwood had the necessary financial and metallurgical knowledge, and Schmitt had the imagination and energy needed to pick the best prospects.

They first turned to southeastern Arizona, at the old Esperanza property, on the other side of Twin Buttes from the Pima. Calumet and Arizona had run some tunnels under the

Above, Duval does some drilling on its Sierrita ore body in 1964; the Esperanza dumps show at left. Below is Duval's Esperanza open pit, seen from the northeast with shops in foreground, as of November, 1969. *(–Both, Duval Corp.)*

copper-stained Esperanza hill in 1906 and 1907, but found only material that was evidently so lean it was not assayed. Harrison Schmitt thought that under 1956 conditions a lot of it might be ore and planned a drilling campaign that by 1957 had developed fifty million tons of ore averaging 0.66 per cent copper and a little molybdenum. The material could be mined by open pit with a low ratio of waste to ore. Atwood directed thorough tests, which proved that a good recovery could be made by flotation; concentrates could be sent to a custom smelter. If the estimates were correct, there would be an excellent profit above the $20,000,000 needed to bring the property to production. The "young engineer," now no longer young, who had started New Cornelia told United Gas they could safely spend the money to build a 10,000 ton per day mill. United Gas hesitated a bit, but finally guaranteed the loans to Duval. Construction was completed in record time, and production started in 1959. Esperanza made money from the start.

The Duval team craved for more. First they took over a great iron-stained mountain at Ithaca Peak, in the northwestern corner of Arizona. Several large companies, including Calumet and Arizona, had tried this area earlier but had found nothing worth-while under conditions in the 1920s. Duval thought they could make it go in the '60s. Harrison Schmitt laid out a drilling campaign on a mountainside so steep that niches had to be blasted out to make room for the drill. The ore was uneven in grade, with enough oxidized copper to complicate concentration, but there was a greater tonnage of somewhat richer ore than at Esperanza. Here too there was no room for miscalculation: the old requirement that a mine must be rich enough to stand champagne and bad management would not apply. As at Esperanza there were no serious errors. United Gas guaranteed loans sufficient to build another 12,000 ton per day plant. The Ithaca Peak or Mineral Park mill started up in 1964 and by 1966 exceeded the Esperanza production.

The Esperanza, lower, and Sierrita, upper, are shown from 18,000 feet with the Sierrita Mountains at top. *(—Duval Corp.)*

Copper Canyon and Battle Mountain, two neighboring districts in northern Nevada, were the next to attract Duval. Two of the greatest copper companies had tackled these showings but found the tonnage too small and the ore too refractory to be worth large-scale production. This failure did not scare off Morris, Atwood and Schmitt; they developed about twenty-four million tons of low-grade open-pit ore in each of the two mines. The possible profit was small. However, they built a 3,000 ton per day mill at Copper Canyon and a leaching plant at Battle Mountain. Production started in 1967, and a lot of metallurgical problems had to be solved, but the Nevada properties of Duval are now a moderate success.

The big one was still to come for the Duval trio. A couple of miles from the Esperanza, in the foothills of the Sierrita Mountains, there was another faintly iron- and copper-stained area. Drilling indicated an enormous body of lean copper- and molybdenum-bearing material that most authorities called waste. Duval would not agree that four hundred million tons of open-pit material containing seven pounds of copper per ton wasn't ore, and the United States Government thought it needed the copper. In 1967 the General Services Administration advanced $83,000,000 against future production; a group of bankers loaned $49,000,000 and Duval, with the help of United Gas, put up the rest of the $156,000,000 required for a 60,000 ton per day plant. It was a big bite for Duval, but it proved a digestible one. The mill started in 1970, and in 1971 it had reached the projected capacity of more than 50,000 tons of copper per year. The Sierrita Mine is one of the great ones and has proved that material carrying seven pounds of copper per ton is ore.

The Duval record was too tempting to be left alone. A much greater industrial corporation, Pennzoil United, decided to acquire both United Gas and Duval; their stockholders resisted, but Pennzoil was too strong. It first bought control of United Gas and consolidated United with Pennzoil. There was still some Duval stock outstanding, and for some reason

Above is Duval's Mineral Park mine and mill near Kingman, Arizona. Below, some mining activity is under way at the Battle Mountain property in September, 1969. *(—Both, Duval Corp.)*

— probably connected with taxes — Pennzoil wanted to get practically all of the Duval stock. In 1968 the price of Duval went up to more than $400 per share on the American Stock Exchange (it had been $68 in 1967); this was a windfall for the Duval stockholders who had held on.

Since the Pennzoil takeover and after the death of Harrison Schmitt, Duval has continued to hunt for new mines, but with no outstanding results.

OTHER LOW-GRADE ARIZONA MINES

The spectacular success of Pima, Mission and the Duval copper mines led to one of the most active exploration campaigns in the history of copper. The big mining companies and many oil companies established offices in Tucson; the desert was crawling with engineers, geologists and geophysical crews. The once sleepy, half-Mexican town of Tucson had become the copper hub of the world.

As in the past, the hunt was successful. Anaconda got the best of the new mines, and Jack Knaebel probably deserves as much credit as anyone. He had developed the uranium mines of Anaconda in New Mexico and later refused to stay in New York as a high administrative officer because he liked the West.

Many exploring companies had toyed with the idea that there might be another "big one" under the gravel slope west of the Santa Cruz River, southeast of the sulphide lenses of the old Twin Buttes mine and northeast from the Esperanza. Successors to the old Twin Buttes Company owned part of the area and wanted a lot of money for it. A. B. Bowman, president of the Banner Mining Company, finally met their terms and located a lot of additional claims. He could afford to take a chance after his success at Mineral Hill.

Bowman drilled a few blind holes through the rapidly thickening desert wash between the foothills and the river. He found material that might be ore if there was enough of it, but Bowman did not have money enough to prove whether or not the theory was correct. He was ready to sell to a bigger company. Anaconda was the best bet, since it was about to lose most or all of Chuquicamata and its other great Chilean mines by expropriation. Jack Knaebel had the courage and vision to take a chance; so he paid what seemed to others an enormous sum for the Twin Buttes and other Banner properties.

It proved a good speculation. Within a couple of years Anaconda had developed 290,000,000 tons of ore that averaged 0.88 per cent copper. It was open-pit ore, but the thickness of gravel and barren capping that must be removed before production could start reached a depth of more than three hundred feet. Anaconda must move a hundred million tons of barren material before it could produce a pound of copper at Twin Buttes. The engineers developed new methods and equipment; fast belt conveyors carried the waste to great dumps, which were landscaped to avoid spoiling the looks of the desert. Anaconda spent nearly $200,000,000 in stripping and plant construction before it could start the mill.

The Twin Buttes operation has been plagued by misfortunes ever since the mill started, in 1971. An unexpectedly large proportion of the copper was in oxidized form, and mill recovery was disappointing. The tonnage treated was lower than expected, and costs were higher. To add to the troubles, the management had made no firm commitments for treatment or sale of the enormous tonnage of concentrates. It had been assumed that much of them could be sold in Japan, but due to the slowing of world demand for copper in 1971 and 1972, Japan could not take even the copper it had definitely contracted for. Anti-pollution laws caused a reduction in the amount of concentrates the United States could accept. As a result unsold concentrates piled up at Twin Buttes. The operating loss became staggering. Several of the leading Anaconda executives were discharged, apparently partly because of the expropriation of the Chilean mines and partly because of the losses at Twin Buttes. The new administration announced that productive capacity at Twin Buttes had to be doubled before the profit would be satisfactory. Someday the enormous

Twin Buttes mine will be an outstanding success, but thus far it has been a sad fiasco.

MAGMA MINES

Another great technical accomplishment was the San Manuel Mine of Magma Copper, which resulted from intelligent geological deduction. Dr. B. S. Butler, of the U. S. Geological Survey and the University of Arizona, deserves most of the credit. Before and during World War II the Government was hungry for more copper, and Butler remembered a little red hill on the road from Oracle to Mammoth, Arizona, where there was copper stain with the iron. Old workings had proved that the red hill was of no value, but a great area east from the hill was covered by gravel and barren "Gila Conglomerate." A couple of miles north a succession of adventurous owners had with varying success mined lenses of a complex ore: they had to save gold, silver, lead, zinc, molybdenum and vanadium in order to make a profit at this Mammoth Mine. Butler wondered what lay under the gravel and conglomerate south of the Mammoth and east of the red hill. It was a dubious gamble, but the need for copper made the Bureau of Mines accept Butler's recommendation to drill a few holes. Three of these holes found copper-bearing rock – maybe it was ore.

As the area was privately owned, the Bureau could do no more exploration. It was up to private initiative, and the owners wanted so much cash that most companies backed away. John Gustafson, geologist at the Superior mine of Magma Copper, was more optimistic and convinced Magma and its principal owner, the Newmont Mining Company, that the chance was worth taking. Magma drilled dozens of deep holes through the gravel and conglomerate. As the holes found ore, shafts and underground workings followed. The result was about four hundred million tons averaging 0.75 per cent copper. This deposit, called the San Manuel, was the greatest success that has resulted from many decades of costly and

The Magma smelter at Superior, Arizona, is shown above in 1952, Below are the headframes and plant of Magma's underground mine, San Manuel. *(—Both, Newmont Mining Corp.)*

intelligent study by the Geological Survey and the Bureau of Mines all over the country.

There were lots of problems. The cover of barren material was so thick that mining must be underground, and the rock was shattered by faults so that extraction drifts below the "caving" areas had to be supported by reinforced concrete. Another complication came from the fact that Anaconda had developed a smaller extension of the ore body; neither company could mine the ore within hundreds of feet of the boundary without caving the property of the other. Magma had enough ore without including the part near the sideline and spent a hundred million dollars for underground preparatory work, a 35,000 ton per day mill, smelter, branch railway, town and all needed facilities. Part of the money was advanced by the Government against future deliveries of copper; later the advance was converted to an $80,000,000 loan from banks and insurance companies. The financial load was a heavy one for a moderate-sized company. Fortunately, everything worked well from its start in 1950, and the San Manuel output has increased to more than 100,000 tons of copper per year.

Even this was not all. The top of the San Manuel ore body was a rolling, gently dipping plane; evidently the ore had been cut off by a great fault. A little drilling by various groups above the fault further down the dip had found only lean material, but David Lowell, a Tucson consulting geologist, thought there was still a chance. The Quintana Petroleum Corporation – a moderate-sized oil producer – wanted to find a mine and hired Lowell to help. Lowell thought of the area southeast of the San Manuel: from the wavy form of the top of the great ore body, as shown in published articles, he worked out the probable direction of the movement. Quintana secured the ground that might contain the ore at a price that seemed to others unconscionable. Lowell planned the drilling; the holes had to pass through more than two thousand feet of worthless material before they found the ore, but the theory

The San Manuel smelter and plant were quite developed by 1971. Below is the interior of the cell house at this refinery. *(–Both, Newmont Mining Corp.)*

proved to be correct. The Quintana drilling indicated over 560,000,000 tons of ore that averaged 0.7 per cent sulphide copper in this Kalamazoo ore body.

The task of bringing the deep, lean ore to production was beyond Quintana's financial or technical resources, but Magma could do it for much less money. In 1968 Magma, soon to be wholly owned by the Newmont Mining Company, bought the Kalamazoo for $27,000,000. San Manuel is increasing its plant capacity to 60,000 tons of ore per day to take care of the new ore. When full capacity is reached, Magma, and through it Newmont, will be the fourth largest copper producer in the United States.

While small compared with the San Manuel and Kalamazoo, the old Superior mine of Magma has added to the success. In the sixty years since the Gunn-Thompson syndicate had found a spectacularly rich vein of copper on the edge of the desert sixty miles east of Phoenix, the Magma Company that they formed was a good, moderate-sized success. As the vein was not wide they had to run long drifts and sink deep shafts to keep ahead of their mill and smelter. And they had to fight excessive underground heat: at 4,500 feet depth the rock temperature was 145 degrees. Mining was barely possible, even with costly air-cooling plants.

As the old mine was approaching the end, exploratory drifts to the east, under the rugged cliffs of Apache Leap, found a new ore occurrence. This rich ore replaced the twenty-degree dipping limestone beds in the south wall of the Magma vein. The temperature was not as high as in the lower levels of the vein itself. By 1971 Magma had developed in the new ore body ten million tons that averaged 5.8 per cent copper.

Magma had acquired a great group of claims further south, in which a succession of optimists had spent millions in trying to find ore bodies under outcrops of veins that looked, on the surface, like the Magma. They had found just enough ore to lead them on to spend more money. They realized that the ore might be far out to the east, under Apache Leap, but they

The Berkeley open pit at Butte is shown above in 1963; it now covers more than twice this area. Below is the smelter at Inspiration in 1972, with excavations for a new smelter that will meet air-pollution control requirements. *(–Above, Anaconda Report; below, Inspiration Consolidated Copper Co.)*

could not afford to spend the added cost of exploring this area with deep shafts and drifts thousands of feet long. From its new ore body Magma can do this at its leisure, and may find other bonanzas like the ten million tons it has found east of the old mine.

An episode in the early exploration of the ground south of the Magma throws light on the characters in the mining world sixty years ago. In intervals between visits to Ajo, that young geologist was looking for other mines for Calumet and Arizona. Some claims south of Superior were offered to the company by Hugh Daggs; the young fellow liked them, and Hugh became a good friend of his. Daggs had been leader of a violent sailors' strike in the Pacific and all sorts of other things that conservatives thought were not respectable. Hugh's father had been one of the locators of the ground near Globe that became the Miami Copper Company. He and one of his partners, "Rimrock" Thompson, had moved from Globe to Superior, and in an argument over their interests Thompson had killed the elder Daggs. A jury called it self-defense. While Calumet and Arizona was talking over an option with Hugh, who had inherited his father's interest, Black Jack Newman met Hugh and asked, "Is it true that you have thrown in with Rimrock to give an option to C. and A.?" Hugh said, "Yes, that's right!" Black Jack protested, "But don't you know that Rimrock killed your daddy over those claims?" Hugh replied, "Yes, but I reckon the old s-- of a b---- had it coming."

BUTTE LOW-GRADE OPEN-PIT MINE

As the great Butte mines became deeper, costs increased until many of the mines had to be shut down except in times of high copper prices. The end of the Richest Hill on Earth seemed to be approaching.

Fortunately Anaconda's geological department had sampled the lean material on comparatively shallow levels between the old high-grade stopes. Much of it might yield a profit with cheap open-pit mining. The company started a small open pit in the "Horsetail" area under the low ground southeast of Butte Hill. This was so successful that after running and sampling thousands of additional feet of workings the management decided to shift operations to a great pit that would remove much of the Hill itself, with its old mines, as well as the lower area to the southeast. Acquisition of all of the claims in this great area made the change possible. The flotation plant at Butte has been expanded to produce more than a hundred tons of copper a day from 10,000 tons of ore. More than a hundred million tons of one per cent ore are estimated. In periods of high metal prices the deep, rich underground mines will add to the production. The dramatic change is being made gradually, with no long shutdown of the pit.

Government limitation of the amount of sulphur smelters can discharge into the atmosphere threatened the Butte and Anaconda operation. To avoid this Anaconda announced in February, 1973, that it is building a plant to treat all of the Butte concentrates by leaching with ammoniacal solutions. Construction and operating costs will be lower than for smelting. This is a great triumph for the Anaconda metallurgists. It may mean the end of all copper smelting.

WHITE PINE MINE, NORTHERN MICHIGAN

The White Pine Mine of the Copper Range Company is too important to overlook. It is not in the "fashion" of the low-grade porphyry copper deposits, but it belongs to the same period and is low grade.

This was not a new discovery; some of the boulders of native copper found in northern Michigan by early explorers were near what is now White Pine. The richer Calumet ore bodies outshone the outcrops sixty miles further southwest, and the several small, copper-bearing lodes in lava, like those near Calumet, did not amount to much. In the sandstone and shale closer to the lake a rich little silver deposit was found, but this also was only a "teaser." South of the silver mine prospectors found native copper in a shale bed that sloped about twenty degrees east. Early attempts to mine this bedded ore failed.

Five miles south of the old silver mine there was a richer copper showing. The beds here were offset by the large White Pine Fault. Irregular bodies of native copper ore came to the surface in the crushed material of the fault. Since the world was hungry for copper during World War I, Calumet and Hecla engineers thought they might increase their production by mining the fault ore. They built a small old-fashioned concentrating mill and in three years mined 900,000 tons of ore, from which they recovered fifteen pounds of copper per ton. When the price of copper dropped after 1920, Calumet and Hecla shut down their White Pine Mine.

Before the shutdown, C. and H. geologists had recognized that in addition to the ore in the fault zone, copper occurred in flat beds in the shale north of the fault. They drilled a few holes to test these beds, but the thickness of the copper-bearing shale was only a few feet and assays were low; milling tests gave an extremely poor recovery. Calumet and Hecla saw no possible value in this thin bedded deposit.

Mr. C. H. Benedict, the distinguished chief metallurgist for Calumet and Hecla, later told the author that they had failed to recognize the importance of the chalcocite, a copper sulphide, that occurred with the native copper at White Pine. They had assumed that, as in other mines in northern Michigan, chalcocite was hardly more than a curiosity; they made no attempt to save this mineral. This assumption cost C. and H. a great mine.

William H. Schacht, manager of the old Baltic mine of the Copper Range Company, had no such belief. He needed a new mine to replace the aging Baltic, and when he heard of the bedded material northeast of the White Pine Fault he had vision enough to think it might be valuable. Morris F. LaCroix, president of Copper Range, was just as ambitious and farseeing. Together they made one of the best teams in the story of copper.

In January, 1929, Copper Range bought the White Pine Mine at "public sale" for $119,000. Copper Range did not have much money, and for years development at White Pine was slow, but they kept on drilling. As all of the holes gave remarkably uniform results they increased the interval between holes from five hundred feet to one or two thousand feet. The greater interval proved reliable in showing the average thickness and copper content of the ore. Year by year the White Pine bedded ore body continued to grow; by 1945 it totalled two hundred million tons, averaging twenty-two pounds copper per ton.

The area covered by the ore body soon amounted to many square miles. Donald E. White, Michigan representative of the United States Geological Survey, and Jack Rand, geologist for the Copper Range Company, found out that nearly all of the copper in this great deposit was in two beds of 2 per cent or 3 per cent ore, each less than five feet thick, separated by from five to fifteen feet of almost barren sandstone and shale. Copper was chiefly in the form of extremely small specks of chalcocite. The deposit was like that in the Mansfield Copper Shale in East Germany, which had been an important source of copper for centuries, but White Pine was even greater.

The most serious problems at White Pine were, first, how to save this copper; and second, how to mine the ore economically with the sort of equipment used in flatly dipping coal mines. Both were tough ones. One of the best research companies in the country could recover little more than half of the minute grains of chalcocite; the Copper Range metallurgist, making tests in the Freda mill that treated Baltic ore, raised this to 80 per cent. White Pine was possible. The Copper Range staff and several consulting mining engineers worked out a method of mining the ore economically, but the minimum thickness the great machines could dig was eight and a half feet. This meant that the 2 per cent or 3 per cent ore in the actual ore-bearing beds must be "diluted" in mining to little more than 1 per cent copper; however, cheap mining would make this profitable.

Bill Schacht died before his dream came true, and in 1945 the Bill Schacht Shaft was named in his honor. Experimental mining and milling of ore from this shaft removed the worst of the uncertainties.

As soon as success seemed assured, Morris LaCroix and his staff planned to build a moderate-sized plant to solve remaining questions – but the government wanted more copper quickly. After long negotiations, Copper Range made a contract under which the Minerals Procurement Agency loaned Copper Range over $57,000,000 and agreed to buy 275,000 tons of the copper output. Construction of the 10,000 ton per day plant began in 1952, and production started in 1955. In 1956 the White Pine operation produced 73,000,000 pounds of copper and made a profit of $9,000,000.

Morris LaCroix died before the plant was running. Under the later leadership, first of "Pete" Lally and then of James Boyd, the plant was gradually enlarged, and the capacity is now more than 150,000,000 pounds of copper a year. A life of several decades is assured, since White Pine probably has the greatest reserves in recoverable copper of any mine in North America, and it is still growing. In copper output and reserves it dwarfs Calumet and Hecla and all of the other Michigan copper mines.

The depression of 1971 and 1972 resulted in a drop in the price of copper. At the same time the general inflation of wages and other costs and the increasing distance of working faces at White Pine down the wavy dip from the surface caused costs to increase. In 1972 White Pine and its parent, Copper Range Company, suffered a serious loss. But the country and the world will continue to need the copper from this great deposit. In the long run a readjustment of the price of copper to conform with the value of the dollar will again make the White Pine one of the leading profitable copper producers.

OTHER LOW-GRADE UNITED STATES DEPOSITS

Several other great low-grade copper ore bodies have been found in the United States. In the Arizona desert south of Casa Grande, El Paso Natural Gas took over for unpaid bills the Lake Shore Mine. A small company had tried a new recovery process for the partly oxidized ore at the Lake Shore but failed. El Paso Natural Gas carried on extensive drilling and found hundreds of millions of tons of ore; more than four hundred million tons averaging 0.7 per cent may be mined by open pit. It is partly oxidized and will require a process that has never been used on a large scale. Seventy million tons of deeper 1.6 per cent ore has its copper in sulphide form; the richer inclined deposit is comparatively thin and mining will be difficult.

The problems at Lake Shore were so serious that El Paso Natural Gas asked many companies to bid for a chance to take over control. The Hecla Mining Company won out and is now sinking deep shafts and preparing to mine both types of ore. Hecla hopes to start large-scale production in 1974. The Lake Shore is one of the toughest challenges that have faced copper miners, but it should become one of the greatest mines.

Even more difficult troubles must be solved at Safford, Arizona. Here Kennecott, Phelps Dodge and American Metal-Climax all have hundreds of millions of tons of 0.7 per cent or 0.8 per cent copper ore, covered by a great thickness of barren rock, and the ore must be mined underground. The presence of part of the copper in oxidized form will make recovery difficult. Kennecott is studying the problem with the Atomic Energy Commission, and hopes to shatter the deep ore body with an atomic blast, then dissolve the copper from the broken rock. They must first prove to their own satisfaction that no radioactivity will escape to pollute either water that may drain into the great irrigated valleys of Arizona or the air above them. Then they must assemble evidence so convincing that even Arizona farmers won't sue. Even with the best technological brains and unlimited money, production from the billions of pounds of copper at Safford seems far away.

Other great lean ore bodies, such as the iron- and copper-bearing material near Yerington, Nevada, seem equally far in the future. In the Copper Creek area across the San Pedro valley from San Manuel, Magma and others have indicated ore bodies that are not good enough to bring to production just yet. The downthrown parts of the Miami and Inspiration ore bodies across the Pinal Fault are known to be there, but the size and mining conditions are unknown. Other copper-bearing areas are even more uncertain. As it takes from five to ten or more years to develop and bring to production a large low-grade copper mine, these copper deposits are for the distant future.

Cates and Lavender proved that three or four-tenths of a per cent copper may make profitable ore. This proof has resulted in new United States mines that contain more than six billion tons of ore. From these new bodies about forty million tons of copper can be recovered. Material that has been changed from waste to ore by technological progress in older mines adds nearly the same amount. In all, forty years' supply at the present rate of domestic consumption is assured. We don't need to worry about our copper for a long time to come.

"Spud" Huestis, left, made Bethlehem Copper a great success and so paved the way for a whole flock of profitable low-grade, open-pit copper mines in Canada. Below, open-pit mining at Craigmont Mines Ltd. is now being replaced by underground operations. *(–Left, Bethlehem Copper Co.; below, Placer Development Ltd.)*

12

New Canadian Low-Grade Mines

CANADA lagged behind the United States in finding great, new low-grade ore bodies. When the lean ore bodies were discovered, they were only an indirect offshoot of the developments Louis Cates and Harry Lavender had started in the United States. In Canada the old-fashioned prospector led the way. New tools for exploration by company and government geologists and technicians found a lot of ore, but they followed rather than led.

BETHLEHEM COPPER

H. H. Huestis, known to everyone in southwestern Canada as "Spud," should get most of the credit. The place was the Highland Valley in southern British Columbia. While this valley is only a few miles from the Kettle Valley branch of the Canadian Pacific and a little farther from Ashcroft on the main line, heavy timber and an almost continuous cover of glacial detritus had slowed down any real search for ore.

Copper had been known in this area for many years. The Geological Survey of Canada had reported "disseminated copper and molybdenum sulphides." The British Columbia Department of Mines drilled a few holes in 1919, and well-known Canadian companies did more drilling. They found widespread copper values and stringers of rich ore, but both the companies and the Survey concluded that the copper was too scattered to be ore. Highland Valley seemed dead by 1927.

Spud Huestis disagreed. He was one of the prospectors who "backpacked" through the wilderness and found many of the Canadian mines by hard, intelligent hunting rather than technological skill. Between prospecting trips he took odd jobs to keep alive.

One of these jobs was on the Guichen Ranch near Merritt. Spud looked over the scattered outcrops northeast of the town and called it a "prospector's dream." There was copper everywhere. While values were spotty, Spud thought the grade might equal that of material that was called ore at Bisbee and Morenci. "All it needed was funds and nerve," he wrote.

Two friends of Spud's, P. M. Reynolds, a chartered accountant, and John A. McLallen, a lumberman, financed the initial development. In 1955 they formed the Bethlehem Copper Corporation and started work. Results were favorable enough that later in the year the American Smelting and Refining Company took an option and drilled many thousand feet of holes. This drilling indicated two ore bodies, each of which contained nearly fifty million tons averaging 0.7 per cent copper. This was not enough for Asarco, which was spending tens of millions of dollars to bring to production richer ore bodies elsewhere. In 1958 Asarco gave up its option, and Bethlehem seemed dead again.

It took Spud Huestis and his partners two years to secure further financing. This time the Sumitomo group of companies in Japan put up the money for a 3,300 ton per day mill and preparations for open-pit mining; construction was rapid. On December 1, 1962, Bethlehem made its first shipment of copper concentrates to Japan. In its first fiscal year Bethlehem produced 25,000,000 pounds of copper from one and a quarter million tons of ore for a net profit of $2,822,564. Since then the mill capacity has been greatly increased. In the eight years since production started, Bethlehem has repaid its debts and paid dividends of about $17,000,000. Many more good years are ahead.

This would have been success enough for most prospectors, but it was only the beginning for Spud. Bethlehem acquired what it called the Valley Copper-Lake area northeast of the original property. Cominco joined Bethlehem in drilling Valley Copper; eighty per cent of the billion tons of 0.5 per cent copper ore are now owned by the great Cominco Ltd., and twenty per cent by Bethlehem, which will treat its share of the ore in its own mill.

Bethlehem has still another great deposit forty miles away at the Maggie Mine, where drilling has indicated a hundred million tons of ore. Still other possible ore bodies are under study by Bethlehem.

OTHER NEW CANADIAN MINES

Spud Huestis' spectacular success has encouraged other companies to try their luck. Prospectors called some of the mineralized areas to the attention of company exploring teams, and careful geological study, aided by all of the new geophysical and geochemical methods, followed. It wasn't romantic, but it was effective. The first to find ore were comparatively small Canadian companies that usually formed stock companies in which the public held a fairly large interest; the result was a Canadian stock-market boom.

The first success was the Craigmont, a few miles south of the Bethlehem. The Placer Development Company, led by John Simpson, found this one with the help of a magnetometer survey. The ore is richer than that at Bethlehem, but not as large. The larger resources of Placer permitted Craigmont to beat Bethlehem in coming to production. For the first few years after production started in 1961, Craigmont mined by open pit 5,000 tons a day of ore that ran nearly 2 per cent copper; the profit in the first year of operation was $11,000,000. With greater depth, in later years the grade of ore has varied from 1.8 to 1 per cent copper, and mining is being changed to underground methods. Craigmont is not one of the greatest copper mines, but it will be a good one for years to come.

Other copper mines have followed. Granby Copper, as its older mines approached an end, obtained Japanese help in financing the Granisle mine, on an island in Babine Lake

In 1937 the author, left, and Mack C. Lake inspected the Glacier Peak Mine in Washington. Drilling at this site is shown below. Plans to equip Glacier Peak were abandoned because of opposition by those who thought scenery enjoyed by a few was more important than copper that would benefit many.

in central British Columbia, where only a few tens of millions of tons of 0.5 per cent ore were known. Inexpensive open-pit mining and good management by Larry Postle, of Granby, have resulted in beautiful profits since the mill started in 1966. The mill is being greatly increased from the original 5,000 tons per day.

More recently the greater Canadian and United States mining companies have taken a more active part in British Columbia copper developments. Noranda developed Brenda Mines Ltd., sixty miles southeast of Craigmont. This is the leanest of all of the successful new mines; a 24,000 ton mill was started in 1970 to treat the 177,000,000 tons of open-pit ore that averages 0.18 per cent copper and 0.049 per cent molybdenum. The grade cannot get much lower.

The Granduc Mine is the last of the Canadian copper mines that has started production. It is not really low-grade; before 1960 there were developed 43,000,000 tons of ore that averages 1.73 per cent copper. But the almost impossible location, high up among the cliffs and glaciers on top of the Coast Range east of Stewart, Alaska, made progress slow and costly. It proved too expensive for Granby, which had taken over the property from a small Vancouver company that had prospected the area by air; Granby leased the property to Newmont and Asarco, who formed the Granduc Operating Company. It took them nearly ten years to drive a ten-mile tunnel at an elevation below the worst of the snow and ice and to build a 7,500 ton per day mill. In all, they had to spend $115,000,000. Strikes and a disastrous snowslide were partly responsible for the delay, but in spite of all the discouragement Newmont, the operator, persisted. Milling on a small scale started in November, 1970. Full production had not been reached in 1972. Granduc is a remarkable tribute to courage, skill and money!

At least seven other great new copper mines are being prepared for production. Four were producing copper in 1972: the Lornex Mining Corporation, controlled by Rio Algom

Lower or Tide camp of the Granduc Mine at Berendon Glacier, British Columbia, is shown above. The beginning of operations at Gibraltar Mines Ltd. in British Columbia in 1972 is below. *(–Above, Newmont Mining Corp.; below, Placer Development Ltd.)*

Mines Ltd., a Canadian steel and uranium producing company; the Highmont Mining Corporation, controlled by the Teck Corporation gold interests; Island Copper, owned by the Utah Minerals Company; and Gibraltar Mines Ltd., a subsidiary of Placer Development. Together these ore bodies have over a billion tons of 0.45 per cent copper ore. The Norman-Bell project of Noranda Mines on Babine Lake and the Similkameen Mining Company, subsidiary of Newmont, are smaller, each with sixty to eighty million tons of 0.5 per cent copper ore. They should start in 1973. Additions to British Columbia production before 1974 should total 265,000 tons of copper a year; the Valley Copper-Bethlehem operation will be at least two years later. Still other British Columbia deposits are under development.

In finding the new British Columbia mines old exploration techniques have been stretched to the utmost and new ones have been devised. Almost unlimited money has been poured into the province. Technological problems have been successfully solved; financial ones have also been solved, at least for the present. But here is the weak point: nearly all of the new low-grade copper mines in Canada, as well as some in other parts of the world, depend to a large degree on Japanese financing. Japan has assumed a future industrial growth in keeping with the fantastic progress of the past decade. Can the boom continue? The contracts with Japanese companies are clear and firm, but Rolls Royce and Lockheed also had contracts that seemed just as firm. No country and no company can meet its obligations when it has expanded too rapidly on borrowed capital. British Columbia and the world will hold their breaths until the growing use of copper outside of Japan catches up with world production.

The Sulfatos copper prospect in Chile, shown in January, 1915, is at 18,000 feet altitude. Its spectacular green staining turned out to be iron sulfate instead of copper. Below, workmen pack out ore from inclined workings in Los Bronces Mine, Chile, at an altitude of 16,000 feet. This mine is part of the present Andina mineralized area. *(—Both, photographs by the author)*

13

Other Parts of the World

OTHER copper-producing parts of the world are poor compared with the United States and Canada, and this is illustrated by their use of copper. The more than a billion people in the "advanced" countries use more than twelve pounds of copper per person per year, and this amount has increased by 15 per cent in the past ten years. The nearly three billion inhabitants, generally nonwhite, of the countries that are hopefully called "developing" consume about a quarter pound of copper per person per year, and there has been no gain in the past decade. Of course the hungry, poorly clad and poorly housed people would like to approach our way of living, but this would require an unbelievable amount of copper and all the other materials that go into making a comfortable life. The demand is there, but the people can't pay for the amenities.

Even more drastic than the poverty of those who need more copper is the attitude of governments in many of the countries where great copper deposits occur. In the past few years the natives in most of these countries have taken over from colonial regimes and have wrested control of their economy from great foreign companies. They need money to fulfill their promises to their people, and raw materials are the most obvious source of money. Most of the new governments lean toward socialism or communism and think the foreign owners of mines are, to use Theodore Roosevelt's phrase, "malefactors of great wealth." The need of their peoples has forced them to take over or "expropriate" the companies that produce raw materials. Mexico showed the way when it expropriated foreign-owned oil companies; Cuba, Peru, Chile, the Philippines, Zambia, and Katanga (now called Zaire) have followed, to greater or lesser degrees. They have taken over all or controlling interests in the copper and other mines for sums that may prove insignificant.

The resulting economic uncertainty makes it difficult to finance costly new copper exploration and equipment; sometimes it is impossible. To this difficulty are added poor transportation, remote and wild terrain, and often a lack of skilled workmen. It is a tribute to the venturesome spirit of mining companies, as well as to the scarcity of the metal, that any great new copper mines are being found and brought to production in the "backward" countries.

For they are being found. New copper deposits elsewhere are not as numerous as those in the United States and Canada, and some of those that have been found have no plans for production because of a lack of money. However, the list of the new foreign copper deposits is impressive.

DEPOSITS IN LATIN AMERICA

Copper deposits found in the past decade outside of the United States and Canada have for the most part come under the same style of very great, low-grade ore bodies and usually depend on large-scale, open-pit mining. Nearly half of these new discoveries since the early 1930s are in Peru and Chile; they are almost all in areas where small, high-grade copper mines had been operated for years or centuries. The newer, low-grade bodies have resulted from patient and costly study and exploration, mostly by companies that already had properties in South America.

The newer mines are for the most part richer than those in the United States and Canada. The first and greatest was the El Salvador mine of Andes Copper, which was owned by Anaconda. By the early 1950s the end of the older Potrerillos mine of Andes was in sight, but Anaconda geologists found a new mine almost on the doorstep of Potrerillos: in the next canyon to the north there was a great area of altered rock that contained streaks of oxidized copper ore. "Vin" Perry and his staff thought the worthless surface material might change in depth to "disseminated" copper ore. Drilling gave results that exceeded their wildest hopes; they found 340,000,000 tons of 1.4 per cent copper ore. The copper content was enough to cover the cost of underground mining. Optimistically Andes changed the name of the deposit from "Indio Muerto," or Dead Indian, to "El Salvador," or the Saviour.

It was that for Andes. Equipment was comparatively inexpensive, as much of the old Andes plant could be used. Production at the rate of 80,000 to 95,000 tons of copper per year started in 1959. Except for the expropriation by the Chilean government, Andes Copper would have had a great future.

Most unusual of the new South American copper mines is the Exotica; as its name indicates, this is a new type of deposit. In many other copper areas there are comparatively small bodies of gravel or conglomerate impregnated with copper that was dissolved and carried down by surface water from eroded parts of ore bodies. On the slope between Chuquicamata and the San Pedro River these "secondary" copper deposits are far from small. Anaconda found the ore by accident as it drilled in areas where it planned to place plants or waste dumps. The drill holes found ore everywhere, covered by only a mantle of barren gravel; the estimated tonnage is now 136,000,000, averaging 1.35 per cent copper. The old leaching plant that treated the shallow oxidized ore at Chuquicamata is well suited to recovering Exotica copper. Anaconda made a contract with the Chilean government under which the government was to receive 25 per cent of the profit after taxes and Anaconda the rest. Copper production started in 1970. Because of operating troubles it had not reached planned capacity in 1972. As Chile is expropriating all of the

Exotica, as well as other large copper properties, the great expected profit will be only a "teaser" for Anaconda.

One more Chilean copper mine is so great that the story of copper would be incomplete without it. This is the Rio Blanco or Andina property near the top of the Andes east of Santiago. Engineers for Anaconda and others — including the author — examined the striking outcrops fifty-five years ago, but the inaccessibility of the district, high up among the snow and ice fields of the Andes, scared everyone away. Chilean owners kept on producing a little rich ore at Los Bronces by primitive methods. After a lot of negotiations with the Chilean government, Cerro de Pasco decided to take a chance; it made an agreement by which the Chilean government was to receive a 30 per cent interest and Cerro 70 per cent. Drilling developed 120,000,000 tons of 1.58 per cent ore, which would have been a bonanza except for the fact that operating at 15,000- to 17,000-foot elevations looked impossible. Cerro solved this problem by running a long tunnel, with the portal below the worst of the snow and ice, and building a mill in underground chambers cut out of the rock; milling is by caving methods. It cost $157,000,000 to bring Andina to production; of this the Chilean government advanced $19,000,000, the Export Import Bank $56,000,000, the Sumitomo Company of Japan $32,000,000, and Cerro the rest. Production started ten months ahead of schedule in 1970; the full capacity of nearly 60,000 tons of copper per year was reached late in 1971. The decision of Chile in July, 1971, to expropriate the Andina, as well as other Chilean copper mines, for considerations determined by the government makes it doubtful if Cerro and the creditors will do much more than get their money back.

Thus far Chile has failed to produce the copper it had promised to deliver to China and other purchasers. No one knows whether or not the government can get technical and administrative ability as well as money enough to operate economically. Of course the big companies hope the government agency will come back to them with its hat in its hand. The result may have a profound effect on nationalization in other backward lands.

The Southern Peru was another great new South American copper mine. Ever since early Spanish times copper had been known in the rugged desert mountains between Arequipa and the coast. Geologists of Cerro de Pasco examined properties five hundred miles south of the old mines and saw a chance of finding great ore bodies. Officers of the company took options from Peruvian owners, and the drilling was encouraging.

The cost of preparing the remote mines for production would be so great that the Cerro officers were in no hurry. They refused to pay the few tens of thousands of dollars one of the owners wanted, as they thought the part ownership they already held would protect them. The American Smelting and Refining Company didn't think so; it bought the share of the recalcitrant owner. A bitter lawsuit followed. No one knew just what one owner could do, under the Peruvian law, if his partners disagreed. Asarco finally won control of two of the great ore bodies — the Toquepala and Quellaveco — while Cerro and its new partner, Newmont, got Cuajone, the third. Together the three contained a billion tons of 1 per cent open-pit ore.

It would have been costly to have separate operations; so in 1954 the three companies transferred their claims to the Southern Peru Copper Corporation. With Phelps Dodge as a fourth partner they spent $150,000,000 constructing a 110-mile railroad from the port of Ilo to the Toquepala Mine, building a smelter, a 30,000 ton per day concentrator and a town, and removing 120,000,000 tons of barren "capping." Most of the money was borrowed from the Export Import Bank.

The companies combined their engineering talents under the leadership of Asarco, and production started in 1960. For the first few years the grade of ore was from 1.7 to 1.5 per cent copper, and profits were spectacular. Even with 1 per cent ore in 1969, Southern Peru earned over $65,000,000, but an increase in Peruvian income tax from 54.5 to 68 per cent in 1970 has reduced earnings. However, Southern Peru has already yielded a great profit over the investment.

Since 1964 Southern Peru has been working on its second great ore body, the Cuajone. Under new Peruvian laws, equipment must proceed rapidly, regardless of what the company thinks is economic. Thus far there are no definite arrangements for securing the $400,000,000 that it will take to bring the Cuajone to production. As long as present laws are in force, it is anyone's guess how much the Cuajone and its sister, the Quellaveco, will be worth to Southern Peru.

The other enormous low-grade copper properties found by American companies in Peru are in a still more dubious state. Not only does the government insist that preparations for production proceed rapidly, but in addition Peru takes a large share in the ownership and restricts exportation of any profits remaining after the heavy income taxes. Under these conditions, the companies cannot raise the money the properties require and have had to turn back to the Peruvian government, without recompense, several of the greatest known copper ore bodies in the world. Asarco has abandoned 590,000,000 tons of 0.75 per cent open-pit ore at Michiquillay; Anaconda will give up 150,000,000 tons of 1 per cent ore at Cerro Verde; and Cerro de Pasco will lose four great new low-grade ore bodies near the rich old Cerro mines. These are in addition to the 200,000,000 tons in the Quellaveco mine of Southern Peru.

The Peruvian government has tried without success to induce foreign governments or companies to equip the new properties, but they all make careful studies and back away. So, for an indefinite time the world loses hundreds of thousands of tons of the copper it needs; the Peruvian workmen lose thousands of good jobs; and the Peruvian government loses hundreds of millions in revenue.

Two other large Latin American low-grade copper properties have been developed in recent years by cooperation between local governments and the United Nations.

The greatest is the Caridad Mine in northern Sonora, Mexico. For more than sixty years examining engineers had been intrigued by widespread iron staining, with streaks of oxidized copper and silver ore, in the rugged mountains southeast of the Pilares Mine of Phelps Dodge. Fly-by-night stock companies failed, and the extremely low grade of the sulphides found many years ago made success seem hopeless.

Both the United Nations and the Mexican authorities were more interested in improving the way of life of the people than in quick profits. The Mexican Geological Survey under Mr. G. P. Salar made a careful geological study near the old Caridad high-grade copper-silver mine. Franc R. Joubin, who had led in the discovery of large uranium deposits in eastern Canada, went to work for the United Nations because he loved exploration. Between 1962 and 1968 the Mexican Survey and the United Nations together found ore that assayed nearly 1 per cent copper. Mexico made a contract with Asarco to continue the work and bring the property to production, with the Mexican government retaining a large interest; Anaconda joined the team later. Now reserves total half a billion tons of open-pit ore that averages nearly 0.8 per cent copper. In spite of the forbidding location, El Caridad will be one of the world's great copper mines.

It will take $240,000,000 to prepare the mine for production, including a railroad, smelter, and town in addition to the mill and mine preparation. Plans call for production at the rate of 75,000 tons of copper per year by 1975, and the skilled engineering talents of Asarco and Anaconda should ensure success. Caridad will add greatly to the resources of Mexico as well as to the funds of the participating companies. It is the best example of intelligent cooperation between government agencies and progressive companies, and the success will

This photo of Charles A. Banks shows him as Canada's munitions and supply representative in London, 1940. Returning to Canada, he became president of Placer Development when it brought in the great Bulolo gold-dredging property in New Guinea. This started the company's spectacular success in copper, lead and zinc as well as gold. For the last few years of his life Banks, then Sir Charles, was Governor General of British Columbia. Below, plant and open pit of the Marcopper mine, controlled by Placer, is shown in 1969. *(—Both, Placer Development Ltd.)*

be due to the hard work of many engineers rather than to the dramatic vision of one or two.

The United Nations, on the recommendation of Franc. Joubin, has cooperated with the government of Panama in drilling other great low-grade copper deposits. The Botija and Petriquilla ore bodies have indicated 375 to 500 million tons of partly oxidized open-pit ore that averages about 0.5 per cent copper plus the molybdenum equivalent. The Panamanian government is negotiating with foreign companies and governments on methods of financing and equipping the mines. Stimulated by the Panamanian ore, companies are studying other large copper showings near the border between Panama and Costa Rica. All of these deposits would have been worthless a few years ago; for possible success they need both modern technology and the control of yellow fever that Colonel Gorgas initiated nearly seventy years ago. These copper deposits may help to bring the backward Latin American countries out of the slough of poverty.

The success of new low-grade copper mines in the Americas naturally led to exploration in other parts of the world. The number of new mines was not as great as in the United States and Canada, but several ore bodies of the first rank were found.

The discovery and equipment of these ore bodies has resulted from thorough geological and geophysical study. The only romance lay in overcoming almost hopeless physical conditions: ore bodies were scattered from the rain-soaked tropical forests to the edge of the arctic, and the governments of many of the countries were at best dubious. We can be proud of the fact that American engineers took a leading part in solving the problems.

PACIFIC ISLANDS AND ASIA

The first of the new low-grade copper mines in remote lands were in the Philippines, where copper had been mined on a small scale at the Atlas and Lepanto mines. During World

The Ertsberg, or "ore mountain," copper mine of Freeport Minerals is in one of the most inaccessible places in the world, the jungle-covered mountains of West Irian, Indonesia, at 11,500 feet. Above is the mine townsite, providing all living accommodations. Below are steel towers of the 5400-foot-long tramways connecting the mine with the concentrator. *(—Both, Freeport Minerals Co.)*

War II the Japanese mined all of the copper they could, although they did not develop the great low-grade ore bodies. After the war engineers took a look at the lean material between the old mines and found that much of it contained more than 0.5 per cent copper. This might be ore. Metallurgical difficulties, mud slides, and almost impossible transportation facilities slowed down the work. The Atlas started its low-grade mill in 1955; by 1969 it had doubled its plant and turned out 50,000 tons of copper, claiming half a billion tons of 0.5 per cent copper ore in reserve. The Lepanto was smaller and slower, but by 1968 it was producing more than 28,000 tons of copper a year.

Marcopper, in a remote area on Marinduque Island, came more slowly. Placer Development, a Canadian company, had made a spectacular success in gold dredging in the wildest part of New Guinea and was not afraid of primitive parts of the tropics. Three imaginative financiers and engineers, Charles Banks, Karl Hoffman and Frank Short, had overcome almost hopeless difficulties in the gold operation. An Australian-born and Harvard-educated geologist, John D. Simpson, became president of Placer and developed Craigmont and other Canadian mines. Beginning in 1956 Placer became interested in Marinduque through an American named Cadwalader, who had been prospecting in the Philippines since 1919. Placer was the partner he needed to help out his Philippine-controlled companies. Heavy rainfall and mud, complex ore and the determination of the Philippine government to get a big share of any profits slowed down the negotiations. Up to 1968 Placer had financed the Marcopper development by loans. Then Placer made a complicated contract with the Philippine government, the original Philippine companies, several banks and a Japanese company: Placer was to receive 40 per cent of the profit of Marcopper; loans were to be repaid; and the Japanese were to buy the copper.

After the contract was signed, Placer constructed a plant at a record rate. With ninety million tons of 0.76 per cent ore

The Hannekam Tunnel, the portal of which is shown, made it possible to operate Freeport's West Irian project despite the steep, trackless Carstenz Mountains. The 21½ x 18½-foot tunnel was driven at record speed, making 52 feet in the best day. *(—Freeport Minerals Co.)*

in reserve, the 15,000 ton per day mill was none too big. The total preproduction cost was more than $40,000,000. Production by Marcopper started in September 1969, and the operation yielded a profit of nearly $15,000,000 in the first year of operation. Capacity is being increased from seventy-five million to a hundred million pounds of copper a year. John Simpson and Placer can be proud of Marcopper.

Other large new copper deposits in the islands of the western Pacific have recently come to production. They are so remote that no one would have thought of them a few years ago. Their rapid development has been due to a combination of American knowledge and daring with the Japanese need for copper and money. Bougainville, on one of the Solomon Islands, has a billion tons of 0.47 per cent open-pit ore; after expenditures of $350,000,000 it started production at the rate of 100,000 tons of copper per year in 1972. The Ertsberg mine, controlled by Freeport Minerals, is on an 11,500-foot tropical mountain in the wilds of West Irian. It is comparatively high-grade, with thirty-three million tons of 2.5 per cent copper ore. Freeport and loaning companies spent $120,000,000. Production started in 1972 and will reach 65,000 tons of copper a year.

The Salar mine in Borneo may be very great but is not as far advanced. These mines and one or two possible additional ones are making us familiar with names that a few years ago meant only cannibals at the far ends of the earth.

The Sar Chesmeh copper deposit in Iran may be as great as those in the Pacific. The United Nations helped to plan and finance its development, and the Iranian government has negotiated with several European and African companies to join in the operation and financing. Thus far there are no definite plans. This failure to finance a great copper deposit shows that there are limits to the return a "backward" country can get from its mineral deposits.

The O'okiep copper mine in South Africa was well established in 1938, when the above picture was taken. While ore reserves have never been extremely great, Newmont's management continues to find new ore bodies and the operation continues to be profitable. The Palabora Mine, also in South Africa, has grown to be very great. The open pit is shown below, with the plant in the background. *(–Both, Newmont Mining Corp.)*

NEW AFRICAN DEPOSITS

Africa also has shared in the copper discoveries since the 1930s. Most of the new ore has been in extensions of the large, comparatively rich mines in Katanga (Zaire) and in Northern Rhodesia, now called Zambia. Production from the older African mines increased by a third from 1960 to 1970, and the change from European to native control in the late 1960s has not, thus far at least, resulted in any drastic decrease in production, save at Mufulira. This greatest of underground copper mines was shut down by a flow of mud and water in 1969; the temporary loss is at the rate of 180,000 tons of copper per year. It has been suggested that the flooding was due to lack of care and maintenance after the Zambian government took control.

Comparatively few great new copper deposits have been found in Africa in the past forty years. The discoveries resulted chiefly from hard work and great expenditures, rather than adventurous prospecting in the wilderness.

By far the largest of the new African copper mines is the Palabora, in the Republic of South Africa. This ore body is of a new type, with magnetic iron oxide and rarer metals associated with the copper. Metallurgy was the great problem, but Newmont, as manager, together with American Metal-Climax and smaller South African companies, solved the difficulties brilliantly. Palabora started production in 1966 with ore that assayed well over 1 per cent copper. The success with the better ore enabled it to reduce the "minimum grade" of minable ore to 0.2 per cent copper, and production has increased to more than a hundred thousand tons of copper per year. As known 0.5 per cent ore totals four hundred million tons, Palabora is one of the best of all the copper mines.

Three post-1955 mines in Zambia would be called great if they were not overshadowed by the four older Zambian operations. These new mines are the Bancroft, Chambishi and Chibuluma, all developed by the older companies. Their combined output of about 110,000 tons of copper per year is less

The Selkirk (left) and Boise shafts at Mufulira are shown above. The surface plant at the newer Chibuluma copper mine in Zambia, shown below, is one of the most modern in the Copper Belt. *(—Both, Roan Consolidated Mines Ltd.)*

than that of any of the original giants: Mufulira, Rhokana, Nchanga, and Roan Antelope. The new mines are extensions of the old Copper Belt.

Two valuable new copper deposits are being developed in the Congo by Japanese capital. Together they have more than a hundred million tons of 2 to 5 per cent ore, but it will be several years before they are in production.

NEW EUROPEAN COPPER DEPOSITS

In Europe the fashion of low-grade copper deposits was slow in taking hold. They are still hardly more than dreams in the minds of the developers, and most of them are partly or wholly government owned.

Majdanpek in Yugoslavia sounds like one of the best and is said to have indicated three hundred million tons of 1 per cent ore; Japanese interests are investigating another possible large area in northern Yugoslavia. The Madarlaga low-grade copper deposit in one of the old copper-producing districts in Turkey is being studied and prepared for production by an American construction and engineering company. The Medet mine in Bulgaria is said to have developed a hundred million tons of 0.5 per cent ore. Production is not in sight.

Rio Tinto has had a complicated history in the past forty years. As in many "poor" countries, the Spanish government was unwilling to have this great resource worked for the benefit of foreign owners. The English company patiently worked out a fair compromise. In 1954 a group of Spanish banks bought the mines and plants of Rio Tinto and put Spanish executives in charge; the English company kept an interest. In 1965 the directors decided to bring to production the "Cerro Colorado" disseminated deposit. They formed a new company in which the English Rio Tinto group and Patino Mining Co. joined other technical and financial companies in bringing this deposit to production. The new plant, which should be completed in 1974, will carry on all stages of a balanced operation from mining to production of refined copper. It will treat 10,000

The Cerro Colorado mine and plant at Rio Tinto, Spain, are shown above. Below is the open pit. *(—Above, photograph by George Argall, editor of* WORLD MINING; *below,* WORLD MINING *and Miller Freeman Publications)*

tons of 0.79 per cent copper ore per day in addition to 5,000 tons of gold- and silver-bearing oxidized material. Forty million tons of copper ore are developed; there may be much more lower-grade material. At the same time another company will continue to mine copper-bearing pyrite. The new plants will assure successful operations for many years.

In the old Outokumpu copper-nickel district in northern Finland recent development has found a great tonnage of possible open-pit material that averages less than 0.2 per cent copper. Success will depend on recovering sulphur and iron as well as copper.

Two or three very low-grade copper deposits in southern Siberia complete the list. We do not know whether or not they would be successful operations if they had to make a profit.

The author and Superintendent H. A. Kursell of the Berezovski Mine, in the Altai Mountains of Siberia, were photographed at the mine by Mrs. Kursell in July, 1917. A few months later Lenin's revolutionaries captured the area, killing many employees, and the couple barely escaped to China and later to the United States. As far as is known, the mine never reopened.

These headframes operate the twin shafts of the Nchanga copper mine in Zambia. The three divisions of Nchanga Consolidated Copper Mines Ltd. now produce nearly half a million tons of copper per year *(—Charter Consolidated Ltd.)*

14

What Next?

THE number of copper deposits of the first magnitude that have been developed since 1930 is impressive. Still more startling is the proportion of the new mines that are in the United States and Canada. The following table shows where the discoveries have been.

COPPER DEPOSITS DEVELOPED SINCE 1930

	No. of Mines	Millions of Tons of Ore	Average Percentage of Copper
United States	24	6,570	0.72%
Canada	21	3,268	0.80
Latin America	12	3,403	0.97
Pacific Islands and Asia	8	2,033	0.71
Europe	9	1,530	0.51
Africa	6	690	1.30
Total	80	17,494	0.80

These new mines contain 140,000,000 tons of copper. Other mines are not well enough developed to permit reliable estimates, and the rate at which discoveries are being made is decreasing. In addition to the new mines, nearly as great a quantity of copper has been found in lower-grade ore surrounding the older mines.

Of course, the number of copper mines "of first magnitude" changes from year to year, and new deposits are still being found. Also, the selection of a possible thirty or forty thousand tons of copper per year as the lower limit for the great mines is an artificial one. The table just shows the approximate magnitude of new discoveries.

The total seems like an awful lot of copper. It would be too much at the maximum pre-Depression consumption of two million tons of copper per year, but world consumption has skyrocketed to over seven million tons in 1970. At this rate the large new mines would supply the demand for twenty years. Increases in reserves at older mines and at small mines may add at most another twenty years. If world consumption continues to increase at the past rate of nearly four per cent per year, total present reserves would last for little more than twenty years. This is far from adequate for an essential metal like copper.

If plans for production from all of the world's new mines were realized, there would be a large excess in the next few years. But actual production never equals productive capacity. Strikes and government restrictions in the "emerging" countries reduce actual production to between eighty and ninety per cent of theoretical capacity. We may have some overproduction for a year or two, but over a longer period we will be lucky if the mines now known can keep pace with the increasing demand. Unless there are startling new discoveries, there is sure to be a serious shortage of copper.

There will probably be startling changes in the copper industry, but in the wrong direction. Concern for the environment is the chief reason. For every ton of copper that is produced, about the same amount of sulphur must somehow be gotten rid of. In the past, most of the sulphur has been discharged into the air in smelter gases, and the resulting pollution has injured crops and possibly health. The smelters are now putting in costly plants to change the sulphur into sulphuric acid, and so keep it out of the air. But there is no possible market for the potential six million tons of acid from copper smelters in the United States — or thirty-four million tons in the world. If the air is not to be hopelessly fouled, much of the sulphur must be fixed in a form in which it will not do any harm. This will mean mountains of waste products.

The legislatures in copper-producing states of this country are almost in a panic. Under pressure from conservationists and farmers, they have passed laws requiring smelters to eliminate ninety per cent of the sulphur from their gases by 1975. This demand shows a fantastic faith in technology: the copper companies must find in what forms the sulphur will be inert, work out a process for changing the smelting gases to those forms and install equipment for making the inert sulphur compounds and storing them. The smelters say it can't be done in the five years; legislators say it must be done. At best the process will require hundreds of millions of dollars, and copper smelters may have to choose between an increased cost—five or ten cents per pound of copper—and shutting down. There will surely be interference with copper production in agricultural areas, such as Arizona, Japan and Europe. If the antipollution groups have their way, the interference will be severe, and it will come soon. It may mean a devastating copper shortage.

Even if there is no such reduction in copper production, the rate at which new copper deposits have been found is inadequate. The past forty years have seen the most successful copper exploration in history, but in spite of this, consumption almost kept up with discoveries. If the 1970 consumption increases at the rate of four per cent, compounded, the recoverable copper found in both old and new mines in the past forty years would be used up in less than twenty-five years.

There are two possibilities. The first is that consumption won't continue to increase; however, with a growing world population this would mean a decrease in the use of copper per person. Copper is so absolutely essential for transportation, production of fertilizers and most of the other requirements of modern life that an end to the growth in copper consumption would mean a decrease in the standard of living. For the advanced parts of the world this would be a major inconvenience; for the backward, already hungry areas it would be a disaster.

The second alternative would be new sources of copper, and these would have to be much greater than those developed in the past forty years. Where are such new copper deposits? No one knows! Discoveries on such a scale must be completely different from those since 1930. Our future copper can no longer come from a reduction of the minimum copper content of ore: we are already working ores down to 0.2 per cent, and there isn't much further to go. The increase cannot result from increased recovery, since recovery is already more than 90 per cent in most great mines. There isn't room for any great further increase.

More low-grade copper deposits like those of the past thirty years will surely be found, but engineers, geologists and geophysicists have already scoured the earth. New mines of the old types will be scarcer and harder to find. If found, they are likely to be extremely low-grade and far from transportation facilities. The cost, per ton of copper, of finding and bringing to production major new deposits is likely to exceed any available source of money.

The present era of new discoveries made by intense scientific study is approaching an end. But ends have been in sight before. New types of ore bodies and new technical achievements may, as so many times in the past, bring new periods of copper discovery. We can't say what or where or when they will be, but when man needs the copper badly enough, he will find it. Copper will still be King of Metals.

Modern African Copper Mining

In most "underdeveloped" countries, when native governments have taken over foreign-owned mines by expropriation or forced purchase the result has been a serious drop in production. Chile is the most striking example. In Zambia, however, the government that replaced the British-dominated regime negotiated fair agreements with the British and American companies that had developed the copper mines. While the government owns a controlling interest, the companies continue to supply technical supervision, keep plants and methods up to date and bring new mines to production. At the same time they have reduced the discrepancy between European and native wages and trained natives to fill any job they can handle. This far-sighted policy has meant increased, instead of decreased, production.

The photographs in this section, supplied by the two great British companies, show the progress of Zambian copper mines under the new regime.

Water flows into one of the six underground "sumps" in the Bancroft Mine. This mine pumps about 60 million gallons of water a day. *(—Charter Consolidated Ltd.)*

Opposite above, a new pump chamber 300 feet long is being excavated in the Bancroft; it will increase capacity to 100 million gallons daily. Below, using a battery of many pumps in line avoids the height and width of larger pumps, which would require costly support of the ground by timbers. *(–Both, Charter Consolidated Ltd.)*

Above, a long-hole drilling machine explores in the Mufulira Mine. These "percussion" drills, using jointed rods, can reach to about fifty feet depth far more cheaply than any other method. *(–Roan Consolidated Mines Ltd.)*

A four-foot-diameter reaming head is attached to the rods of a raise borer at the Luanshya Mine (Roan Antelope). These borers can bore holes up to 120 meters long, vertical or inclined, several times as fast and more cheaply than older methods of raising. *(–Roan Consolidated Mines Ltd.)*

An instructor in the underground training school at Luanshya teaches new miners how to load drill holes for blasting. *(—Roan Consolidated Mines Ltd.)*

A Caterpillar 950 loads Granby-type cars at the Mufulira, above. Below, at the end of shift, underground workers leave the cage at the Luanshya. *(—Both, Roan Consolidated Mines Ltd.)*

This headframe is at the new North Shaft of Nchanga. It was sunk to exploit the upper ore body, which is partly oxidized and could not formerly be treated economically. *(–Charter Consolidated Ltd.)*

The open pit at Nchanga is a very large operation, as the photographs on this page and opposite above show. Opposite below is the Nchanga surface plant, completed for increased production in 1972. In the middle ground are the mill buildings and twin shafts, with part of the open pit at left center. *(—Three photographs, Charter Consolidated Ltd.)*

These 29 grinding mills in the Mufulira concentrator are the final stage in pounding the copper ore to a fine powder that releases the copper-bearing particles from the barren "gang." *(—Roan Consolidated Mines Ltd.)*

Production started at the Bwana M'Kubwa concentrator's flotation plant, above, in May, 1971. The large coal-fired rotary kiln below is used for drying Nchanga concentrates before roasting. *(—Both, Charter Consolidated Ltd.)*

Compressed air is being blown into the converter at the Luanshya smelter to speed burning off of the sulphur. *(—Roan Consolidated Mines Ltd.)*

In the refinery tankhouse of Rhokana Division, Nchanga Consolidated, very thin sheets of pure copper have been electrolytically deposited on insoluble "anodes" and are here being stripped off to serve as starters for the final cathodes of pure electrolytic copper. Below, at the Mufulira Mine, impure copper from the converter is cast as anodes on this revolving wheel; at fixed points the anode is dumped into water to cool and harden, causing the steam. In the tankhouse these anodes will be immersed in acidified water and attached to positive terminals, and the electric flow will deposit pure copper on the starter sheets at the cathodes. *(—Right, Charter Consolidated Ltd.; below, Roan Consolidated Mines Ltd.)*

This is another view of the Nchanga North Shaft head-frame. *(—Charter Consolidated Ltd.)*

Glossary

ADOBE FURNACE. A primitive furnace made of mud bricks, in which ore was smelted in many parts of the world before the advent of machinery and of treatment of ore on a large scale.

AMYGDALOID. A lava bed that contains almond-shaped or "amygdaloidal" gas cavities. In the Lake Superior Copper District many veins follow or replace amygdaloid beds.

ANTIGUA. An ancient Mexican mine that has been abandoned for a century or two.

ARRASTRE. An early Mexican device for grinding ore, consisting of a rotating arm, usually pulled by mules, that drags large stones around a circular floor of paving blocks, crushing a thin layer of ore as they go. Often quicksilver was added to the ore in the arrastre, to help in recovering gold and silver.

ASSESSMENT WORK. The $100 worth of work per year required to keep the title to unpatented mining claims valid.

AZURITE. A beautiful blue carbonate of copper.

BED ROCK. The solid rock that underlies soil or gravel.

BLAST FURNACE. A columnar furnace for melting ore, into which fuel is charged mixed with the ore, and in which combustion is assisted by a blast of air introduced near the bottom of the column.

BONANZA. A spectacularly rich ore body.

BORNITE. A combination of copper with sulphur and iron in one of the richest copper minerals, with a beautifully varied peacock color.

CAGE. An elevator for transporting men and materials up and down a shaft. The grid of iron bars that often enclose the cage is responsible for the name.

CARBONATE. A chemical combination with carbon dioxide. Carbonates of the metals are often found in the upper portions of ore bodies.

CHALCOCITE. The richest combination of copper with sulphur. In its hard, steely form it is called "copper glance."

CHALCOPYRITE. A brass-colored combination of copper with iron and sulphur; not as rich as bornite or chalcocite.

COLLAR. The top of a shaft.

CONCENTRATE. See *Concentration.*

CONCENTRATING TABLE. A concentrating device that makes use of the fact that metal-bearing minerals are usually heavier than the worthless minerals with which they occur. The finely ground ore, mixed with water, is passed in a thin stream over a sloping table that is shaken with a bumping motion that forces the heavy metallic particles toward one end.

CONCENTRATION. An intermediate process in the recovery of metals from ore, by which part of the worthless material is cheaply eliminated, leaving the metals in a comparatively small amount of richer material called "concentrates." The concentrates can then be treated by some more expensive final process, such as smelting.

CONGLOMERATE. A sedimentary rock consisting of water-worn and rounded pebbles or boulders of older rocks cemented by a fine-grained ground mass. Commonly called "pudding-stone."

COPPER GLANCE. The hard, steely form of chalcocite – the richest copper sulphide.

COUSIN JACK. Colloquial name for Cornishmen. The female of the species is called "Cousin Jennie."

CROSSCUT. A passageway in a mine, usually horizontal, that runs across the vein or other geological formation.

DIP. The inclination of a vein or other structure below the horizontal.

DISSEMINATED ORE. Ore in which the metal-bearing particles are sparsely scattered through a rock mass. From the fact that one common type of copper ore body contains scattered copper sulphides that form only a small percentage of the whole mass, these ore bodies are called "disseminated" deposits. As the rock in which the first ore bodies of this sort were found was porphyry, the disseminated copper deposits are often called "porphyry coppers."

DOWSER. One who searches for subterranean supplies of water, ore, etc., by the aid of a divining-rod.

DRIFT. A horizontal passageway in a mine, usually one that follows the vein or other formation.

ENRICHED ORE. Ore in which the original metal content has been increased by the addition of more metal brought down by slowly descending surface water. This water dissolves the metal from the upper portion of the ore bodies, where there has been oxidation, and deposits it again lower down, due to a chemical reaction with sulphides that have not yet been oxidized.

FAULT. A natural displacement of the rock on one side of a plane compared with that on the other. In the course of mountain building the fault motion sometimes amounts to many miles. Such rock displacements often cause earthquakes.

FLOTATION. A process of concentration in which the metallic particles in ore that has been ground and mixed with water are thinly coated with oil, and subjected to a stream of rising air bubbles that carries them to the surface of a tank. There they are skimmed off as a rich "concentrate."

FLOW SHEET. A diagram that shows the various processes used in the treatment of an ore.

GANGUE. The worthless minerals that accompany a metal bearing mineral in an ore.

GLANCE. Any of various ores having a luster which indicates their metallic nature.

GOPHERING. Primitive extraction of ore by digging out rich pockets or streaks, leaving irregular holes like gopher holes.

GRADE. The percentage of metal or other desired material in an ore.

HIGH GRADE. A high metallic content, or ore that has such a content. Colloquial for rich ore.

HOIST. The engine that pulls the ore (or men and materials) out of a mine.

INCLINE. A passageway into a mine, or between different mine workings, that is neither horizontal nor vertical.

LEACHING. Slow dissolving of soluble constituents of an ore by water or solutions that percolate through it. Natural leaching often takes place in the upper portions of ore bodies. Artificial leaching in tanks, by solutions of sulphuric acid and various sulphates, is used in the recovery of copper from certain ores.

LEASERS (more properly lessees). Those who work in a mine on shares instead of for wages, paying part of the proceeds from the ore they dig out to the owner of the mine as "royalty."

LEDGE. A band of rock or zone within which mineral values are found.

LENSES. Deposits of ore that have well defined edges or contacts, (unlike the "disseminated" deposits that have no definite form but fade away into barren rock), but that have an irregular lens shaped form instead of occurring in tabular veins.

LEVEL. All of the mine workings at a certain horizontal elevation. Also the vertical distance below or above the surface or some other given point. For instance the "1500 Level" may mean all of the workings at an elevation 1500 feet below the collar of a shaft, or that depth itself irrespective of any workings.

LODE. An elongated or tabular geological formation within which ore is found. Less specific than a vein, though all veins are lodes.

MALACHITE. The bright green carbonate of copper.

MATTE. A rich artificial sulphide of copper formed as an intermediate product in smelting, between ore and metallic copper.

MESCAL. A potent Mexican drink made by distilling a form of cactus.

METALLURGY. The art of recovering metals from ore.

MILL. A plant for the concentration of ore – by concentrating tables, flotation, or other devices.

MINERAL. A natural element or chemical combination of elements that occurs in the earth.

ORE. A natural association of minerals from which one or more metals may be won, with profit or fair hope of profit.

ORE RESERVES. The amount of ore that can be estimated in a mine.

OUTCROP. The surface exposure of an ore body or of rock in place.
OXIDATION. Combination with oxygen, either by rapid burning or by slow corrosion or rusting through exposure to the air.

PILLAR. Part of an ore body left in place to help support the rock, while the rest of the ore is mined out.
PLACERS. Beds of sand or gravel that contain gold or other metals washed down from the eroded parts of ore bodies.
PORPHYRY. An igneous rock (or a rock that has solidified from a molten state) in which some of the minerals are well crystallized, floating in a finer grained ground mass like raisins in a cake. As the first "disseminated" copper deposits occurred in porphyry, they are often called "porphyry coppers."
PROSPECT. A hole in the ground in which someone is looking for ore – or, to look for ore in a new locality.
PYRITE. The common sulphide of iron.

QUARTZ. Natural silica, one of the commonest of minerals, both in ore bodies and in barren rock.

RAISE. An underground mine opening that leads upward, either vertically or at an incline.
ROAST. To burn or oxidize an ore in order to eliminate the sulphur that is combined with the metals.
RUBY COPPER. A beautiful red natural oxide of copper.

SHAFT. An opening that leads downward into a mine, either vertically or at an incline.
SKIP. A box in which ore is hoisted through a shaft out of a mine.
SKIP TENDER. The man who loads the ore into a skip, and notifies the engineer, by means of a bell or other signal, when to hoist it.
SLAG. The residue from which metals have been removed by a smelter.
SMELTING. A process for recovery of metals from an ore by which the ore is melted in a furnace, allowing the heavy metallic components to sink to the bottom while the lighter, worthless materials – called slag – are poured or skimmed off from the top.
STAMP MILL. A device for crushing and grinding ore, consisting of heavy iron weights or "stamps" dropped at frequent intervals on a thin layer of ore spread out on iron anvils or "dies."
STOPE. An underground excavation from which ore has been removed.

TUNNEL. A horizontal passageway from the surface into a mine.

VEIN. A tabular deposit of ore or of minerals, – with two dimensions many times as great as the third. Veins usually stand at a considerable inclination from the horizontal.

WINZE. A passageway leading downward from a mine opening.

Index

*Denotes picture reference

GREAT COPPER DEPOSITS OF THE WORLD

● KNOWN BEFORE 1935

○ DEVELOPED AFTER 1935

MERCATUR PROJECTION

SIZE OF AREAS FURTHER FROM EQUATOR EXAGGERATED